SolarWinds Orion Network Performance Monitor

An essential guide for installing, implementing, and calibrating SolarWinds Orion NPM

Joe Dissmeyer

[PACKT] enterprise 🟎

PUBLISHING   professional expertise distilled

BIRMINGHAM - MUMBAI

SolarWinds Orion Network Performance Monitor

First published: April 2013

Production Reference: 1120413

Published by Packt Publishing Ltd.
Livery Place
35 Livery Street
Birmingham B3 2PB, UK.

ISBN 978-1-84968-848-2

www.packtpub.com

Cover Image by J. Blaminsky (milak6@wp.pl)

Credits

Author
Joe Dissmeyer

Reviewers
Richard Jones

Dave Shield

Stephen Stack

Acquisition Editor
Andrew Duckworth

Lead Technical Editor
Sweny M. Sukumaran

Technical Editor
Dennis John

Project Coordinator
Anurag Banerjee

Proofreader
Lesley Harrison

Indexer
Hemangini Bari

Graphics
Aditi Gajjar

Production Coordinator
Shantanu Zagade

Cover Work
Shantanu Zagade

About the Author

Joe Dissmeyer has a strong background in enterprise-class software and IT systems which include VMware, Windows Server, Windows Desktop, Exchange Server, and Cisco. He holds multiple IT certifications and has an A.S. degree in Computer Information Science. Joe currently works as part of a team of network engineers for a company in central Florida. Prior to accepting this position, he was working as a Senior Technician for a healthcare provider, a Domain Administrator for a small college, and a Service Desk Specialist for a Fortune 100 company. Joe is well versed in server, desktop, and network administration.

Aside from his full-time job, Joe is a managed service provider for a few small businesses in central Florida where he provides various remote and onsite IT consulting services. He volunteers his knowledge and skills by participating in the Microsoft Answers forums, the Spiceworks IT Professional Community, and frequently posts troubleshooting and tech articles on his website at www.joedissmeyer.com. Joe is an active member of his local community and shares the visual and audio setup responsibilities with his church's tech ministry team every week.

Joe's specialties are the Windows desktop, Windows Server engineering, operating system deployment, network troubleshooting, and network administration. His biggest strength is that he has a deep understanding of how information technology systems work and how they affect a business.

You can contact Joe through his website at www.joedissmeyer.com, or via e-mail at joe@joedissmeyer.com.

Acknowledgement

There are so many people that I want to thank for their support in writing this book. Without their skills, forethought, support, and expertise I would never have been able to write this book on my own.

I would like to thank my beautiful wife Tasha and our three children, Lauren, Cameron, and Jocelyn, my parents Fred and Sandra and step-parents Mitch and Dora. I also wish to thank Carl, Jennifer, and Megan. But most of all, I want to thank Tim, Lisa and Ashley for putting up with me, helping me watch my kids, and helping me find the time to write this book. To everyone, I could not have done this without you — all of you! I love you all!

I also want to thank all of my colleagues and friends. There are too many to name, but I wish to thank; Stacey F., Mason G., Dave "DJ" M., David S., Ernst S., Jim K., Derek M., Steve M, Jacques A, Joe M, Joe P, Rhys R, Debbie W, Brian Z, every member of the SpiceCorps of Central Florida, and the entire Packt Publishing team. I know that I missed a bunch of you, but you know who you are!

For my brothers in Christ; Daniel Hopper, Glenn Stewart, Phillip Kochanski, and Milan Thaker. Thank so much for loaning me equipment and giving me advice on some of the topics in this book. Also, a special shout-out to Dan Williams at Meraki. May God bless you and your families!

But above all, I thank our almighty God. He has given me a gift to share with the world so that I can glorify only Him. None of the knowledge and none of the skills that I have been blessed with come from myself. They all are a gift from God! It is my desire to serve only Him, the Everlasting one, the most High, and the most holy. Amen!

> "...so that they may have the full riches of complete understanding, in order that they may know the mystery of God, namely, Christ, in whom are hidden all the treasures of wisdom and knowledge." — Colossians 2: 2, 3 (NIV)

About the Reviewers

Richard Jones manages the EU regional IT infrastructure for a leading global manufacturing business, and also works closely with the global team to provide best in class systems and support to the business. Richard's certifications and specialties include Network Management Systems, Cisco technology, and VMware.

Dave Shield has worked as part of the technical support team for the Department of Computer Science at the University of Liverpool for more than twenty years. For most of that time, he has also been one of the core developers for the Net-SNMP project helping it grow from a one-man fork to become one of the world's leading open source network management products.

Both of these environments draw heavily on the open source ethos, and typically involve the use of ad-hoc, in-house developed systems. Seeing how things look from the perspective of a commercial software solution has been a fascinating experience, and has helped clarify some of the advantages and limitations of the open source approach.

Stephen Stack has 15 years of industry experience and is currently the Network Team Leader for an International Financial Services company and Application Service Provider (ASP) with offices in more than 50 countries worldwide. His current role includes responsibility for the company's international MPLS-based WAN, client ASP connectivity and data centers, and managing core data center technologies including security, virtualization and network management solutions.

Stephen has extensive experience in the ISP and SME markets also and his certifications include CCNP, CCDP, MCSE, and SolarWinds Certified Professional (SCP) to name a few. He also has a number of SolarWinds NPM, APM, and NCM deployments under his belt.

A keen golfer, Stephen lives in the picturesque village of Ballycotton located in County Cork, Ireland with his wife Orla and son Rían. Stephen's professional profile and be found on LinkedIn at `http://www.linkedin.com/in/ststack/`.

www.PacktPub.com

Support files, eBooks, discount offers and more

You might want to visit www.PacktPub.com for support files and downloads related to your book.

Did you know that Packt offers eBook versions of every book published, with PDF and ePub files available? You can upgrade to the eBook version at www.PacktPub.com and as a print book customer, you are entitled to a discount on the eBook copy. Get in touch with us at service@packtpub.com for more details.

At www.PacktPub.com, you can also read a collection of free technical articles, sign up for a range of free newsletters and receive exclusive discounts and offers on Packt books and eBooks.

PACKTLIB™

http://PacktLib.PacktPub.com

Do you need instant solutions to your IT questions? PacktLib is Packt's online digital book library. Here, you can access, read and search across Packt's entire library of books.

Why Subscribe?

- Fully searchable across every book published by Packt
- Copy and paste, print and bookmark content
- On demand and accessible via web browser

Free Access for Packt account holders

If you have an account with Packt at www.PacktPub.com, you can use this to access PacktLib today and view nine entirely free books. Simply use your login credentials for immediate access.

Instant Updates on New Packt Books

Get notified! Find out when new books are published by following @PacktEnterprise on Twitter, or the *Packt Enterprise* Facebook page.

Table of Contents

Preface

Have you ever had complaints from your customers about poor network performance? What about trying to find out what your bandwidth utilization is from the edge? If you are an IT administrator, I guarantee that you have had these types of tasks before.

I recall a time when I was an IT administrator of a medium-sized business, working at the company headquarters. The business had a data center hosted in Little Rock, Arkansas with more than twenty different branch offices scattered throughout the United States. The data center was the central hub for all network connectivity for the entire organization with each branch office connected to the data center via private MPLS circuits. One day, my team received a call notifying us that one of the remote locations was without Internet access. There was literally no way for me to know why this happened without spending a great deal of time researching the issue. After an hour, we finally found the cause of the problem. The core router died during a lightning storm at that branch office. We were able to call a local technician to connect a spare router at the branch office and get our customers back online but the damage had been done. The total amount of downtime for our customers was four hours which was completely unacceptable for a company that relies on the Internet to perform its work.

In a completely different example, I was working as a network administrator team member at another company where most of our users use a web-based application to perform their jobs. One day, I received an e-mail alert notifying me that our primary Internet link was down. I contacted our ISP who dispatched a technician immediately. As I was working on crafting a notification e-mail message to the company about the situation, I received a call that many of our customers' Internet connections were very slow, their web application was timing out, and they were unable to work. I informed the customer that we were working on the issue and notified the company of the problem. After a short period of time, the ISP technician arrived and resolved the problem. The total customer downtime for this scenario was 30 minutes.

As you no doubt have observed, there are multiple issues with the first scenario. There was no alerting in the event of any type of network failure which limited the IT department to be proactive in such an event. The second scenario shows some of the best and most used features of a network monitoring system. Thanks to the core monitoring features of the monitoring system, I was able to determine the root cause of the problem quickly and have the ISP technician dispatched as soon as possible. Even though my customers experienced a network outage for 30 minutes, I'm sure you would agree that a downtime of 30 minutes is more acceptable than four hours.

SolarWinds Orion Network Performance Monitor is one of these types of monitoring systems and this book is going to discuss many of its features including what Orion NPM actually is, what it does, the technologies behind Orion NPM, and how Orion NPM can help to make your job as an administrator easier.

What is Orion NPM?

Orion Network Performance Monitor is a scalable, easy to use, cost-effective network monitoring system that provides a complete overview of network environments by monitoring performance and availability. Orion NPM enables you to be proactive in detecting, diagnosing, and resolving network issues and outages and has the benefit of supporting hundreds of types of server, OS, and network vendors including Cisco, HP, Microsoft, Linux, Motorola, Brocade, Foundry, and more.

Orion NPM is used by thousands of public and private companies, educational institutions, and government entities and is a well-known product. Here is a list of important features that make Orion NPM stand out in the crowded network monitoring software market:

Logical, useable, customizable, interactive, drill-down (LUCID) interface

The SolarWinds Orion NPM LUCID interface is one of the key features of Orion NPM. It is a browser-based frontend for the entire SolarWinds Orion monitoring system dubbed the "dashboard". Every section of the dashboard is completely customizable. If you do not like viewing the top-level network map module on the Summary home page, it can be moved to a different menu bar or it can be removed entirely. Each module in every menu bar can be customized as well, or custom menu bars can be assigned to specific user accounts. The personalization and dashboard customization options are almost endless!

Uses standard protocols to poll devices and servers

In order to monitor servers and devices, many network monitoring solutions require an administrator to install and configure specialized client software on each server and network device. SolarWinds Orion uses industry-standard protocols that are already built into the software of each server and device, and does not require an administrator to install any additional software.

ConnectNow topology mapping

One of the most time consuming tasks of a network administrator is the need to diagram the topology of a network. The most common tool used to map out a network is Microsoft Visio, but diagramming a network in Visio can take a great deal of time to perfect. Using the Network Sonar Wizard, Orion NPM uses proprietary "ConnectNow" technology to discover device relationships and automatically map those relationships for you in the Orion Network Atlas.

Mobile views

A simple view of the Orion NPM dashboard can be accessed from a mobile web device's web browser. No special "apps" need to be downloaded and installed from an app store in order to view the Orion Dashboard. Simply navigate to the dashboard URL on your mobile device's browser to view!

Microsoft Active Directory integration

User account authentication can be tied in with Microsoft Active Directory. Single accounts can be added to the account authentication in Orion NPM, or entire Active Directory security groups. This allows administrators to continue to centralize and secure authentication and accessibility on the network.

Role-based access

SolarWinds Orion NPM has a robust access control system that can be as granular as you need it to be. An administrator can grant a variety of permissions to specific areas of the Orion Dashboard, or even administrative portions of Orion NPM. Even more granularity is enabled when role-based access is combined with the integration of Microsoft Active Directory.

Automated network discovery

SolarWinds Orion NPM can be configured to automatically scan your network on a regular basis for devices and servers and add them to the Orion dashboard for monitoring. This helps to get Orion NPM set up quickly for new installations as well as making device management easier for administrators in existing installations.

Multi-vendor device support, universal polling, and custom MIB creation

Thanks to Orion NPM using industry-standard polling protocols, thousands of manufacturers and vendors are supported in Orion NPM. Orion NPM can also import customized MIBs from various vendors.

Conditional group dependencies

Devices and/or servers can be grouped together with defined dependencies in a parent/child relationship. When the parent device is down, only a single alert notification will be sent instead of one for every child dependency.

Wireless polling

Orion NPM can monitor wireless access points and keep historical data of SSIDs, client IP addresses, IP addresses, signal strength, channel usage, and more.

Virtual server monitoring

You do not need to purchase additional licensing just to keep an eye on your VMware virtual server hosts. Orion NPM can do this out of the box! Both virtual server hosts and resident virtual machines for VMware ESX and ESXi are supported.

Report Writer

Orion NPM includes several preconfigured reports. Using the included Report Writer, you can write your own customized reports as well as automate their creation.

VSAN summary

SolarWinds Orion NPM can not only monitor your critical network devices and servers, but also your fiber channel and virtual storage. Orion NPM can alert administrators if VSAN storage volumes have low disk space, low I/O performance, and more. You can drill down to the nitty-gritty details on the fiber channel interfaces including transmitted and received data as well as utilization information.

Community content exchange

SolarWinds has created a comprehensive support community built around the Orion product line called Thwack. You can find expert advice forums, submit feature requests, download administrative scripts and Orion add-ons, free tools, and other content in the Thwack community.

Cisco EnergyWise monitoring

Orion NPM can take advantage of Cisco's EnergyWise software component in Cisco Catalyst switches. EnergyWise is a part of Cisco's "Green Initiative" that monitors power consumption in Catalyst switches that can generate reports and alerts for power-related incidents. For example, if you have a port with Power Over Ethernet (PoE) capabilities and that port has PoE enabled, but the PoE is not in use on that port, Orion NPM can generate an alert for this port. EnergyWise is designed to help IT departments become "more green" and help with reducing power consumption, which will effectively help to lower costs.

Do-it-yourself deployment

You don't need to be an expert to install and set up SolarWinds Orion NPM on your network and you don't need to hire a specialized consultant to do it for you. Orion NPM is designed to be easy to install and set up. It is possible to set up a full Orion NPM solution within an hour! I should know, I've done it myself.

As you can see, there are several core features of SolarWinds Orion NPM that helps differentiate it from the competition. As you become more familiar with Orion NPM using this book, you will discover even more features not listed above!

How Orion NPM monitors your network

The Orion NPM system is a database-driven web application which operates on top of Microsoft .NET server technologies. Microsoft Internet Information Services (IIS) is the web service for the Orion Dashboard and Microsoft SQL Server is the database backend for all information gathered from network devices and servers.

Devices are added to the Orion NPM database either manually by IP address or DNS name, or automatically by using the Network Sonar Wizard. Once a device has been added to Orion NPM, it is polled for data by Orion NPM on a predefined timer, or counter. An internal process consistently runs in the background on the Orion NPM server that checks when to "kick off" the polling engine depending on the time set for a device in the counter. When that time has been reached, the device is polled.

SolarWinds Orion NPM does not poll all devices at the exact same time at a set, predetermined, fixed time. Orion NPM only polls the device when the counter has been reached. It may be difficult to understand this, so here is an example. Imagine that you have a very large network with 5,000 network devices. If SolarWinds Orion NPM was configured to poll all 5,000 devices at precisely the same time, this would act just like a denial-of-service attack and literally take your network down! The counter process is a fantastic feature since it guarantees that Orion NPM won't flood your network with polling traffic and won't cut off your users' network access.

Orion NPM monitors a network using industry-standard protocols to poll data from network devices on a regular basis. The protocols used by Orion NPM to gather network information are Simple Network Management Protocol (SNMP), Windows Management Instrumentation (WMI), Internet Control Message Protocol (ICMP), and Syslog. Depending on the device, Orion NPM will use an appropriate protocol to gather information. For gathering data from a Cisco switch, Orion NPM would use SNMP or ICMP. To gather data from a Windows server, it may use WMI. The following diagram is a simple example of how Orion NPM monitors a network and how that information is presented:

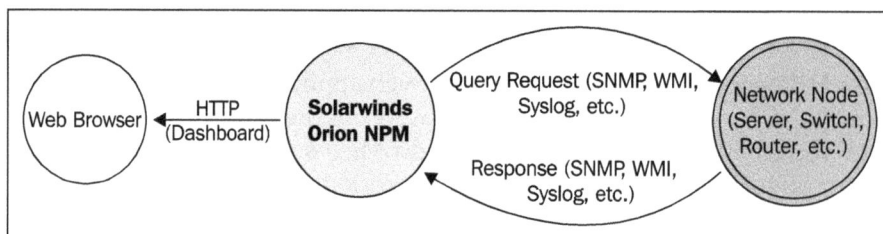

It is important to understand not only how Orion NPM operates, but also understand the technologies, standards, and protocols that it uses. The next few sections describe several standard network monitoring protocols and how Orion NPM uses them.

Simple Network Management Protocol (SNMP)

SNMP is the most commonly used protocol for gathering monitoring data from computer systems and network devices and it consists of three components: managed devices, agents, and network management systems. A managed device could be a switch, router, server, or any other type of network device that has an SNMP agent. An SNMP agent is software on a device that translates data to SNMP-compatible language for transmission across a network to a network management system, such as SolarWinds Orion NPM. SNMP has been around almost since the beginning of the modern computer age and has gone through several revisions.

SNMP is an IETF-standardized protocol and operates in one of two ways; the manager/agent model, and traps. In the manager/agent model, an SNMP agent is configured on a device to allow SNMP communication between itself and an SNMP manager. The SNMP manager periodically grabs the device's information from the SNMP agent. SNMP can gather an endless list of information from a network device such as memory usage, CPU utilization, power supply usage, syslog messages, humidity sensors, and so on.

Most SNMP traffic is initiated by the SNMP manager, but SNMP traps can be configured on an SNMP agent to directly alert the management system of some type of abnormality, such as high CPU usage in a server or maxed-out bandwidth usage from an interface in a router. The information an SNMP trap transmits to alert an SNMP manager of a problem depends on what is defined in its Management Information Base (MIB). Some vendors offer a utility to create custom MIBs for SNMP agents for a particular device.

Orion NPM can use all three iterations of the SNMP protocol; Version 1, Version 2c, and Version 3. Versions 1 and 2c are still considered the de-facto standards of SNMP by many and follow a simple community-based way of authentication using a defined IP port, community string, and/or a read/write community string. SNMPv3 builds on SNMPv2 and offers more robust security options.

SNMP agents are typically disabled by default and must be configured manually by an administrator. The best thing about SNMP is that it is found in virtually every single manageable network device and operating system on the planet so it makes sense that Orion NPM would utilize SNMP extensively.

Windows Management Instrumentation (WMI)

WMI is a management framework built into all modern Windows operating systems which grants administrative visibility to almost every aspect of the Windows OS. Management applications or administrative scripts can be created to view or manipulate components of Windows using WMI in a variety of programming languages. The most common type of administrative scripts that take advantage of WMI are VBScript and Windows PowerShell. Applications such as SolarWinds Orion NPM can make programmatic WMI calls to a Windows computer to access direct information about the operating system such as its IP address, MAC address, SNMP information, event logs, active and non-active services, and more. WMI can gather the same type of information from a computer that an SNMP agent can. Microsoft has a built-in security model for WMI, so before you go querying data from a Windows computer you need to make sure you have the proper access on that computer to do so.

Internet Control Message Protocol (ICMP)

Internet Control Message Protocol is more affectionately referred to as ICMP and it is one of the core protocols of the TCP/IP suite. ICMP allows network devices to send errors, control information, and informational messages to and from network device. PING may be the most commonly used command-line tool in most operating systems that best showcases the ICMP protocol.

Syslog

Syslog is another IETF-standardized protocol for event notification messages. It allows a network device to send event logs and event notifications to an event collection system, usually called a Syslog server or Syslog collector. Almost every network device and network server has its own internal logging system. Using syslog, it is possible to have a device automatically forward its event logs across the network to a Syslog server. Orion NPM has its own built-in Syslog server and stores retrieved syslog messages in its SQL Server database.

What this book covers

This book strictly covers SolarWinds' flagship product, Orion Network Performance Monitor. Inside you will find all of the essential information required to install, set up, calibrate, and administer Orion NPM.

Chapter 1, Installation, tells you how to install Orion NPM.

Chapter 2, Orion NPM Configuration, builds upon the previous chapter and covers the initial configuration of Orion NPM.

Chapter 3, Device Management, discusses how to add devices to Orion NPM, various polling methods, and how to managing devices.

Chapter 4, Network Monitoring Essentials, gives an overview of the Orion website, discusses monitoring routers, switches, and wireless controllers.

Chapter 5, Network Monitoring II, continues upon the previous chapter by discussing server and virtualization monitoring, including universal device pollers.

Chapter 6, Setting Up and Creating Alerts, discusses the alerts and notification system in Orion NPM.

Chapter 7, Producing Reports and Network Mapping, takes a look at the reporting system and network mapping utilities in Orion NPM.

Chapter 8, Maintenance, discusses the various tools and tasks associated with maintaining an Orion NPM system.

Appendix A, Documentation and Support, shows you the online resources you can refer to for more information and support.

Appendix B, The Thwack Community, introduces you to the Thwack Community, a fully featured IT professional community for SolarWinds products.

Appendix C, Additional SolarWinds Orion Software, talks about additional SolarWinds Orion products that can be used to extend Orion NPM's core functionality.

What you need for this book

It is highly recommended to have the following hardware and software available in order to follow along with many of the examples discussed in this book:

- A computer with a 64-bit processor running Windows Server 2008 R2

- A computer running Windows, Linux, or Max OS X with a modern web browser (that is, Google Chrome, Firefox, and so on)

- Microsoft SQL Server 2008 R2 Express Edition

- SolarWinds Orion NPM 30-day evaluation

- A modern Linux OS (that is, Ubuntu 12.04 LTS, Fedora 18, and so on)

- VMware ESXi 4.0 or newer

- A Wireless Access Point and/or a Wireless Controller

- An enterprise-class managed switch (that is, Cisco Catalyst series, Brocade FastIron, HP Procurve, and so on)

- Managed router (that is, Cisco 2800 series, Juniper J-series, Vyatta Virtual Appliance, and so on)
- Managed firewall (Cisco PIX or ASA series, Palo Alto PA-series, and so on)

Who this book is for

This book is targeted to IT administrators that want a quick start to setting up Orion NPM. However, for those that just purchased SolarWinds Orion NPM (or are building a case for their IT Management team to purchase it), this book will assist you with that endeavor. For those that are already using Orion NPM in a test lab or a real-world production environment, this book could be used as a reference training manual. Another reason you purchased this book is because you are already using Orion NPM in a limited fashion and you want to know what additional features are available. One way or another, this book will suit your needs for everything Orion NPM.

Conventions

In this book, you will find a number of styles of text that distinguish between different kinds of information. Here are some examples of these styles, and an explanation of their meaning.

Code words in text, database table names, folder names, filenames, file extensions, pathnames, dummy URLs, user input, and Twitter handles are shown as follows: "For medium to large network sizes, a more appropriate view option is to set the first level to Location then level two to Department."

Any command-line input or output is written as follows:

```
net start SolarWindsTrapService
```

New terms and **important words** are shown in bold. Words that you see on the screen, in menus or dialog boxes for example, appear in the text like this: "Click on **External Websites** in Orion Web Administration and then click on the **ADD** button."

> Warnings or important notes appear in a box like this.

> Tips and tricks appear like this.

Reader feedback

Feedback from our readers is always welcome. Let us know what you think about this book—what you liked or may have disliked. Reader feedback is important for us to develop titles that you really get the most out of.

To send us general feedback, simply send an e-mail to feedback@packtpub.com, and mention the book title via the subject of your message.

If there is a topic that you have expertise in and you are interested in either writing or contributing to a book, see our author guide on www.packtpub.com/authors.

Customer support

Now that you are the proud owner of a Packt book, we have a number of things to help you to get the most from your purchase.

Errata

Although we have taken every care to ensure the accuracy of our content, mistakes do happen. If you find a mistake in one of our books—maybe a mistake in the text or the code—we would be grateful if you would report this to us. By doing so, you can save other readers from frustration and help us improve subsequent versions of this book. If you find any errata, please report them by visiting http://www.packtpub.com/submit-errata, selecting your book, clicking on the **errata submission form** link, and entering the details of your errata. Once your errata are verified, your submission will be accepted and the errata will be uploaded on our website, or added to any list of existing errata, under the Errata section of that title. Any existing errata can be viewed by selecting your title from http://www.packtpub.com/support.

Piracy

Piracy of copyright material on the Internet is an ongoing problem across all media. At Packt, we take the protection of our copyright and licenses very seriously. If you come across any illegal copies of our works, in any form, on the Internet, please provide us with the location address or website name immediately so that we can pursue a remedy.

Please contact us at copyright@packtpub.com with a link to the suspected pirated material.

We appreciate your help in protecting our authors, and our ability to bring you valuable content.

Questions

You can contact us at `questions@packtpub.com` if you are having a problem with any aspect of the book, and we will do our best to address it.

1
Installation

This is where things start to get interesting, so hold on to your hats because we have a lot of ground to cover and only one chapter to do it in! This chapter will help you with both planning and installing your SolarWinds Orion NPM implementation by discussing its system requirements, Microsoft SQL Server configuration, and Windows Server configuration.

By the end of this chapter, you will have learned about the following:

- Orion NPM system requirements
- Windows Server configuration
- Microsoft SQL Server setup
- Configuring Windows Firewall settings for Orion NPM
- Installation of Orion NPM

System requirements

SolarWinds Orion NPM is a Windows-only product and it requires two things; a computer running a compatible Windows Server OS and a Windows computer running Microsoft SQL Server 2008 or newer. In order to plan out how or where you are going to install Orion NPM, you need to know Orion NPM's hardware, operating system, software, and SQL Server requirements. Depending on the Windows Server hardware requirements, you can mix and match most modern Windows Server operating systems and SQL Server editions.

> The latest SolarWinds Orion NPM system requirements information can be found in the Orion NPM Quick Start Guide at http://www.solarwinds.com/documentation/orion/orionDoc.aspx.

Minimum Windows Server hardware requirements:

- 2.0 GHz Intel or AMD Dual Core Processor
- 3 GB of memory
- 2.5 GB of free hard disk space

Recommended Windows Server hardware:

- 3.0 GHz Intel or AMD Quad Core Processor
- 8 GB of memory
- 20 GB of free hard disk space

Minimum Windows Server OS requirements:

- Windows Server 2008 with SP2 (x86 or x64)

Recommended Windows Server OS:

- Windows Server 2008 R2 with SP1 (x64)
- Windows Server 2012 (x64)

Additional software requirements:

- .NET Framework 3.5
- .NET Framework 4.0
- .NET Framework 4.5
- Internet Information Services 6.0 or higher (32-bit mode only)

Supported SQL Server editions:

- Microsoft SQL Server 2008 with SP2 Express, Standard, or Enterprise
- Microsoft SQL Server 2008 R2 with SP1 Express, Standard, or Enterprise
- Microsoft SQL Server 2012 Express, Standard, or Enterprise

Support for Windows Server 2012 was added in the 10.4.0 release of Orion NPM and SQL Server 2012 support was added in the 10.3.1 release. Keep in mind that Microsoft Will drop support for Windows Server 2003 in 2014 with SQL Server 2005 support being dropped soon afterwards. This is why I am not including them in the preceding list even though they are on the official list of supported software. Both I and SolarWinds highly recommend using Windows Server 2008 R2 and SQL Server 2008 R2 at a minimum.

SolarWinds supports Orion NPM installations on virtual machines. However, the following rules apply:

- Microsoft SQL Server must be installed on its own dedicated physical server. Orion databases that reside in a SQL Server instance running inside a virtual machine is not supported.

- Orion NPM installations on domain controllers are not supported.

- Orion NPM virtual machines are only supported in VMware and Microsoft Hyper-V environments. Other virtual hosts (such as Citrix XenServer or KVM) are not supported.

- The virtual machine where Orion NPM is installed must have its own dedicated physical network interface card on the virtual host.

Windows Server setup

Orion NPM runs in native 32-bit mode and can only be installed on Windows workgroup servers or on Windows servers joined to a domain. It is recommended to start with a clean Windows Server installation for best results. If you are repurposing a Windows Server that was previously in use, or if you are planning on using a Windows Server currently in limited use, remove the Internet Information Services role from it. Also, it is not possible to install Orion NPM on a Domain Controller.

Internet Information Services (IIS)

Internet Information Services (IIS) is Microsoft's web server software that is installed as a role in Windows Server. IIS will be enabled and configured by the Orion NPM installer without any intervention or manual configuration needed by an administrator after the installation has completed. The only catch is that the IIS role must not be installed or enabled on the target Windows Server.

It is best to remove the IIS role from the Windows Server using the Server Manager application if it is enabled, since Orion NPM will replace the default IIS website with its own. Do this so that the Orion NPM installer is able to automatically configure IIS correctly. If the IIS role remains active, it is highly possible that the Orion NPM installation will fail.

Microsoft SQL Server

Although it is not required, SolarWinds recommends installing Orion NPM and SQL Server on two different servers on different physical hardware for performance reasons. This recommendation is especially true if you are running SolarWinds and SQL Server in two different virtual machines on the same virtual server host, or if you plan on monitoring more than 100 nodes. Consider the fact that the Microsoft IIS service, the Orion NPM polling engines, and SQL Server processes are incredibly resource-intensive. Installing SQL Server and Orion NPM on the same server, even with less than 3 GB of RAM and a slower processor, might work fine for very small networks (less than 50 monitored nodes), but it is still not recommended to do so.

Orion NPM includes tools that will help you move the Orion database to a different Microsoft SQL Server instance if the need arises. This allows you to initially start with a single Windows Server with both Orion NPM and SQL Server installation on the same machine, then move the database at a later time or when performance becomes an issue.

The official stance from SolarWinds is that SQL Server Express editions are supported, but SQL Server Standard or Enterprise editions are highly recommended. The reason being SQL Server Express editions have been handicapped by Microsoft. SQL Express cannot utilize more than 1 GB of system RAM on the Windows Server where it is installed. This is true regardless of how much RAM is available on the server. SQL Server Standard editions can use up to 16 GB of available system RAM and Enterprise editions can utilize as much RAM is available to it on the Windows server. Also, SQL Server Express technical support is only available by Microsoft through the TechNet and community support forums while the Standard and Enterprise editions are supported under regular support contracts.

With all of this in mind, SolarWinds ultimately leaves it up to the administrator to decide if they want to use a limited edition or an officially-supported business edition of SQL Server. Selecting an appropriate SQL Server edition depends on how many nodes you intend on monitoring with Orion NPM, as well as what you consider an acceptable risk regarding technical support. So the question remains, "Which SQL Server edition should I choose?" Based on my personal experience, you should use the following:

- Windows Server 2008 R2 for the base OS
- SQL Server 2008 R2 with SP1 Standard edition or newer for the database server
- Ensure the Windows and SQL servers are members of an Active Directory domain

SQL Server setup

This section will demonstrate installing and configuring SQL Server 2008 R2 Express with SP1 x64 on a Windows Server 2008 R2 with SP1 computer. The installation and configuration process is exactly the same for the standard and enterprise editions of SQL Server 2008, SQL Server 2008 R2, and SQL Server 2012. Because SQL Server needs to be configured for Orion NPM further than a simple default installation, I will be installing the edition of SQL Server which includes the management tools.

The installation and configuration process for SQL Server is straightforward. The following is what we will be covering in this section:

- Downloading and installing the SQL Server Express software including the SQL Server Management Studio tools
- Enabling the SQL Server services using the SQL Server Configuration Manager
- Verifying SQL Server authentication settings
- Configuring Windows Firewall with Advanced Services to allow traffic to/from SQL Server and the Orion NPM server

SQL Server 2008 R2 Express is available for download for free at `http://www.microsoft.com/en-us/download/details.aspx?id=26729`.

Using the preceding link, download one of the following files from the SQL Server 2008 R2 Express download page:

- `SQLEXPRWT_x64.exe` (329.9 MB) for 64 bit Windows Server
- `SQLEXPRWT_x86.exe` (302 MB) for 32 bit Windows Server

Both of these downloads include the SQL Server Management Studio tools needed to configure SQL Server services after the installation has completed.

Service accounts

Before starting the SQL Server installation, there is a little bit of prep work that needs to be done regarding user accounts and administrative rights on the target server on which you are installing the Microsoft SQL Server software.

There is a point in the SQL Server installation wizard where you must define a user account that has administrative rights. This account will be used by Windows to manage the SQL services on the computer. User accounts that are used exclusively by software to perform internal tasks are called **service accounts**. Service accounts can be created on the local Windows Server, or they can be an Active Directory domain account. If SQL Server is installed on a member server in an Active Directory domain, create a domain service account and assign it administrative rights to the target Windows Server where you will be installing SQL Server. Otherwise, create a local service account on the target server and assign it administrative rights. All that truly matters is that the service account has administrative rights on the target server.

> If you already have a SQL Server 2008 or 2012 instance on your network and do not need to install or set up a new server, please skip to the *Verifying SQL Server authentication settings* section.

Installing SQL Server 2008 R2 Express

Perform the following steps to set up a new SQL Server 2008 R2 Express installation from scratch.

> The installation steps displayed in this section are tailored for SQL Server 2008 R2 Express edition. However, these do apply to the all business editions of SQL Server 2008, SQL Server 2008 R2, and SQL Server 2012 after step 3.

1. Log into the Windows Server computer with the SQL Server service account.
2. Double-click the `SQLEXPRWT_x64.exe` or `SQLEXPRWT_x86.exe` file to launch the SQL Server Express installer.
3. The installer will search for any prerequisite software needed in order to complete the installation. Click on **OK** to continue.

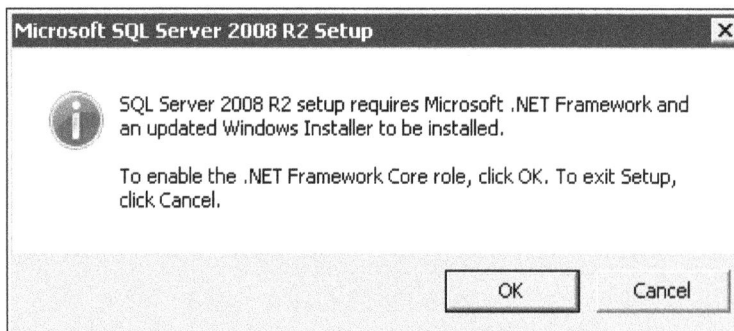

Microsoft SQL Server 2008 R2 Setup [X]

SQL Server 2008 R2 setup requires Microsoft .NET Framework and an updated Windows Installer to be installed.

To enable the .NET Framework Core role, click OK. To exit Setup, click Cancel.

OK   Cancel

4. After a few moments, the **SQL Server Installation Center** will appear. Click on the **New installation or add features to an existing installation** link. SQL Server will automatically install several support files then continue with the installation.

5. At the **License Terms** screen, place a check mark next to **I accept the license terms** and click on **Next**. The setup wizard will automatically install several prerequisite files needed to continue the installation.

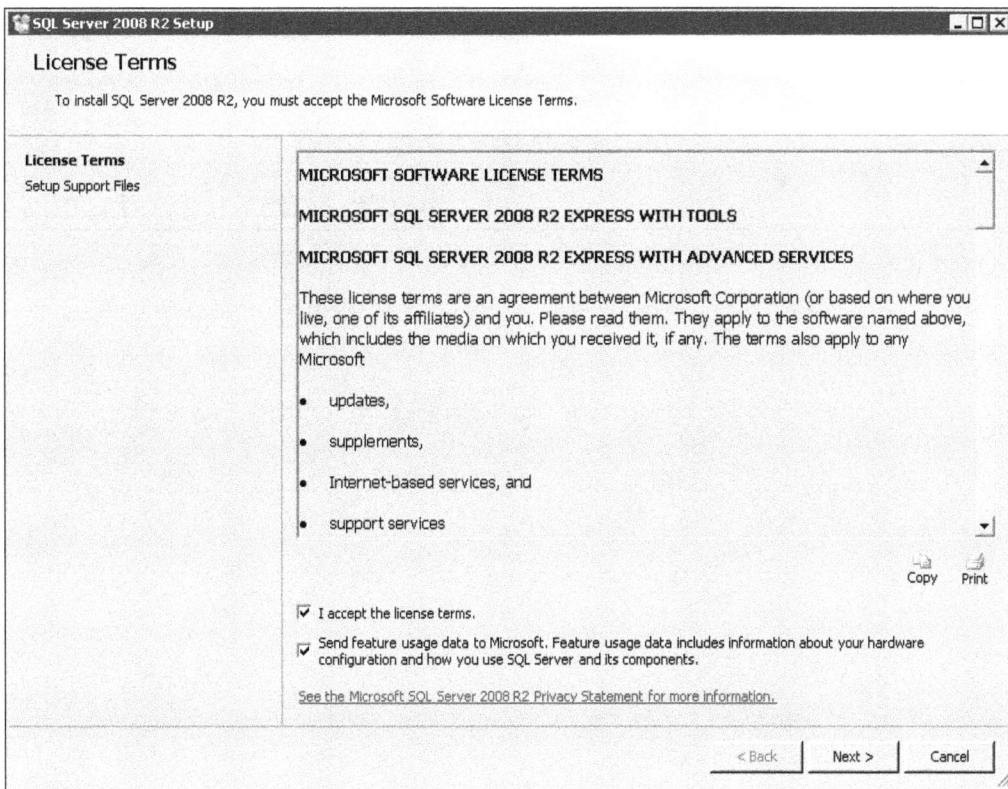

6. At the **Feature Selection** screen, place a check mark next to **Database Engine Services**, **SQL Server Replication**, and **Management Tools – Basic**. Click on **Next**.

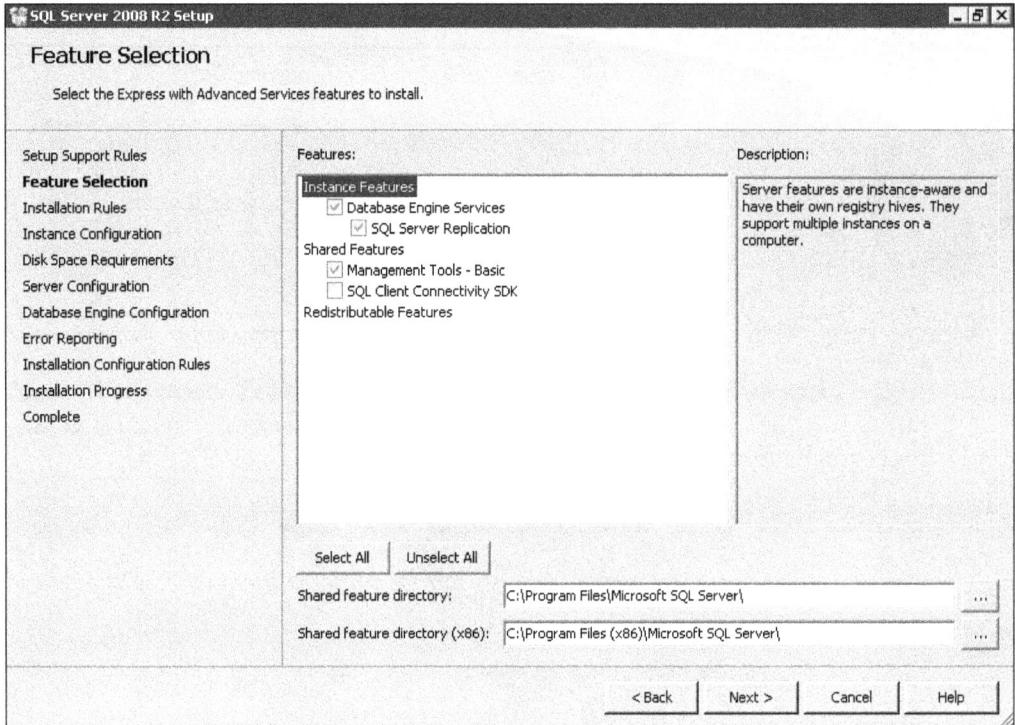

7. On the **Instance Configuration** screen, choose the **Named instance** option and type a name for your SQL Server database instance. For this example, I will use the default name **SQLExpress**. Click on **Next** to continue.

8. At the **Server Configuration** screen, click on the **Service Accounts** tab. It is recommended to use the same user account for both the **SQL Server Database Engine** and the **SQL Server Browser** services. Click on **Use the same account for all SQL Server services** button and define the user account with administrative rights to this computer.

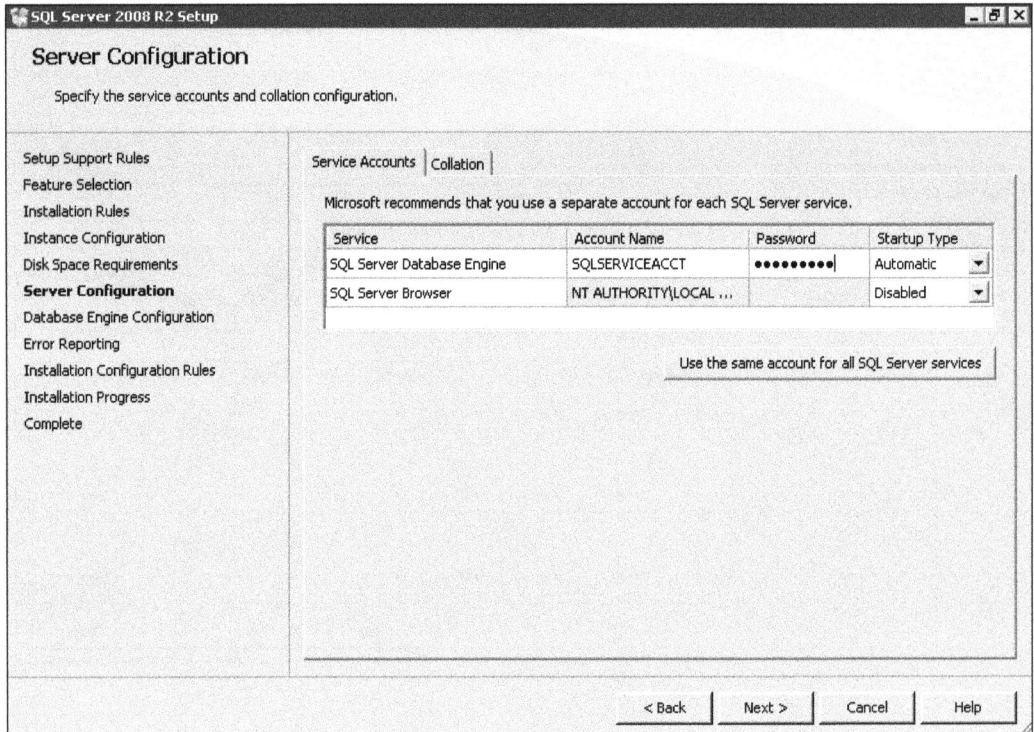

Service	Account Name	Password	Startup Type
SQL Server Database Engine	SQLSERVICEACCT	●●●●●●●●●	Automatic
SQL Server Browser	NT AUTHORITY\LOCAL …		Disabled

9. Click on the **Collation** tab. Verify that **SQL_Latin1_General_CP1_CI_AS** is set for the database engine. Click on **Next** to continue.

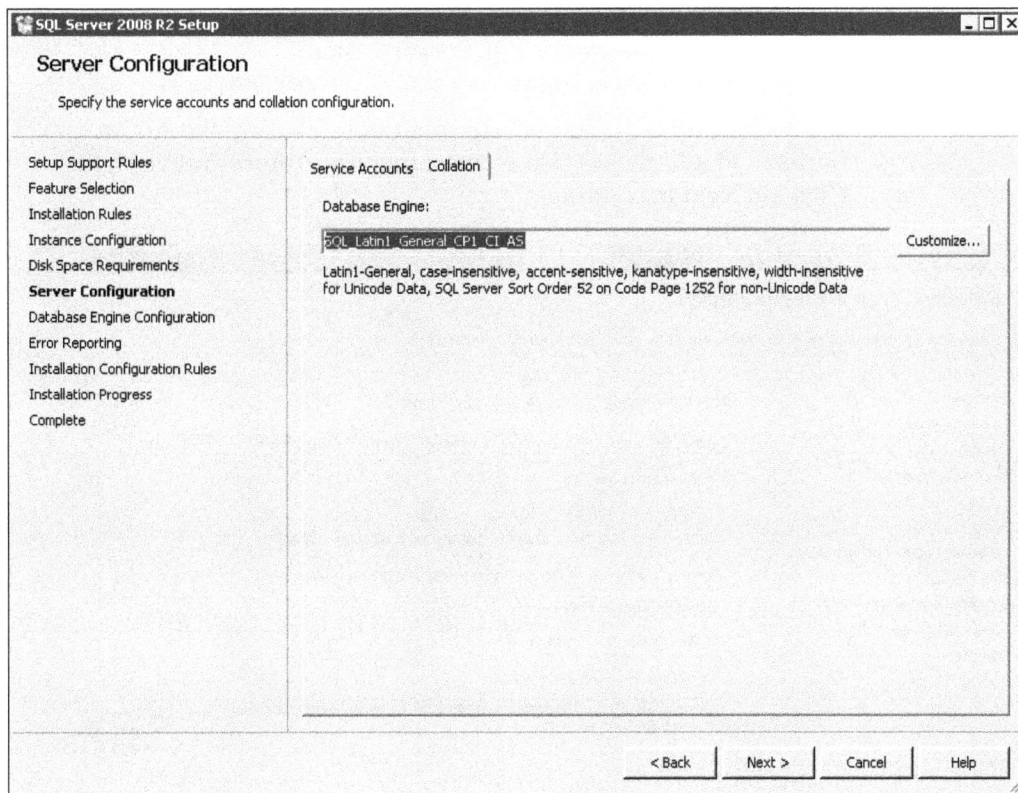

10. At the **Database Engine Configuration** screen, click on the **Account Provisioning** tab and configure the following options:

    1. Under **Authentication Mode**, choose **Mixed Mode**.

    2. Define a password for the SQL Server System Administrator (SA) account.

    3. Under **Specify SQL Server administrators**, click on the **Add...** button and add the service account that has administrative rights. Click on **Next** to continue.

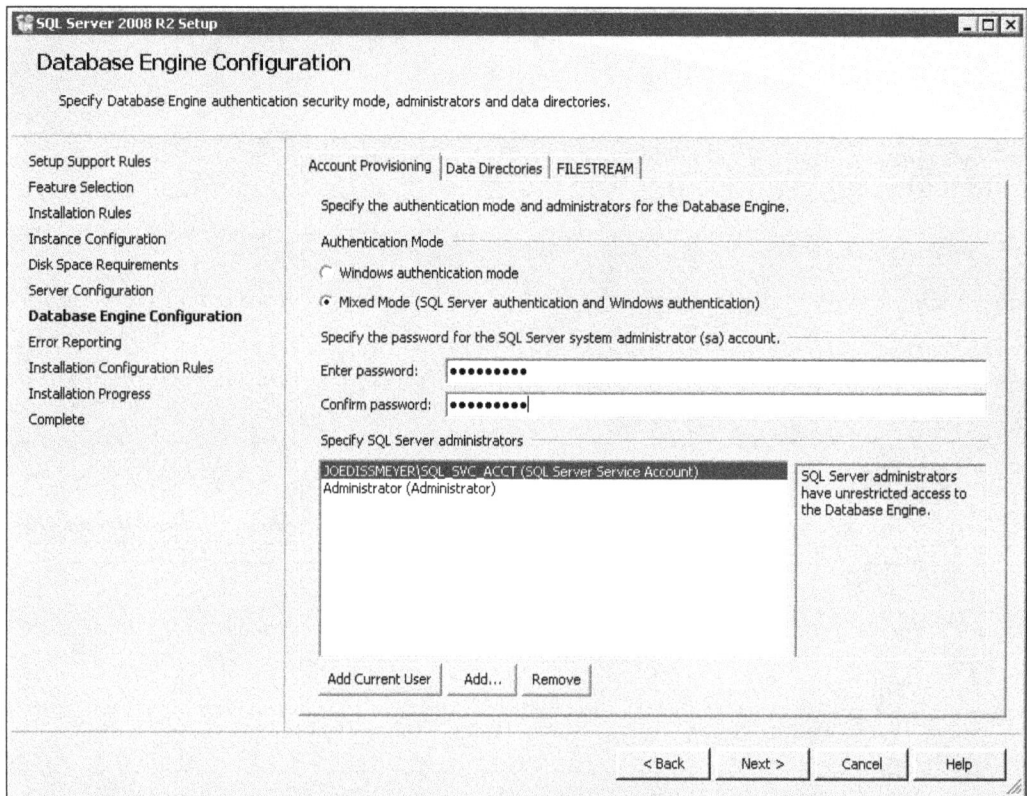

I recommend not to remove the local administrator account from the **SQL Server administrators** group. If the SQL service account becomes deactivated or is locked out due to too many invalid login attempts, you may not be able to access the SQL database.

11. At the **Error Reporting** screen, click on **Next** to start the installation.

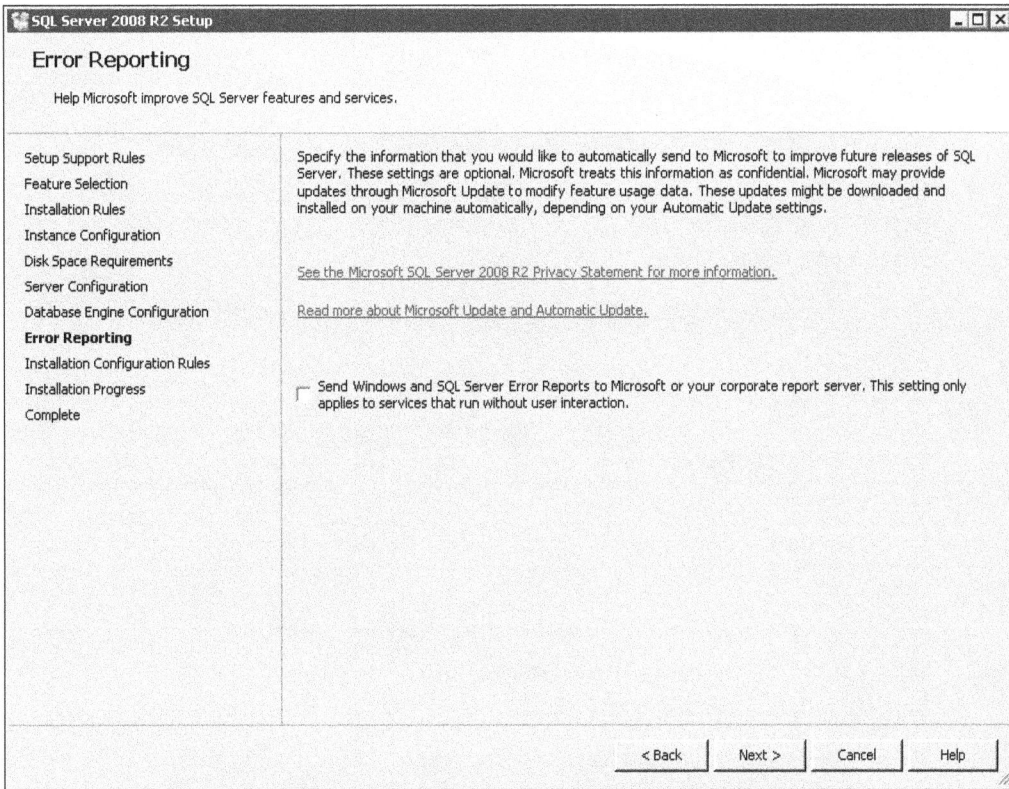

12. The **Installation Progress** screen will now be displayed and start the SQL Server installation process. When finished, click on the **Close** button, then close the **SQL Server Installation Center** window.

Configuring the SQL Server services

Now that the SQL Server installation has finished, we can start the disabled SQL Server services and verify that the SQL Server authentication is configured correctly for SolarWinds Orion NPM.

The following steps are only for those who installed SQL Server from scratch using the instructions from the previous section. However, if you already have a SQL Server on your network and you will be using it to create a new Orion NPM database in it, use the following steps to verify the permissions and configuration for Orion NPM.

1. Launch **SQL Server Configuration Manager** from the Start Menu.

2. On the left-hand side pane, click on **SQL Server Services**. Notice the **SQL Server Browser** and **SQL Server Agent (INSTANCE NAME)** are both stopped.

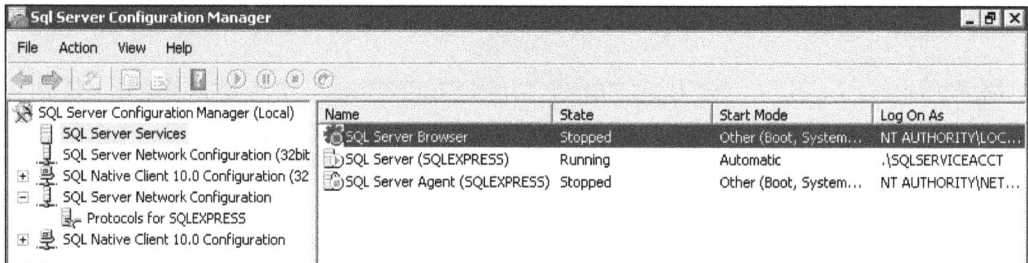

3. Right-click on **SQL Server Browser** and choose **Properties**.

4. The **SQL Server Browser Properties** window appears. Click on the **Service** tab, change the **Start Mode** from **Disabled** to **Automatic**.

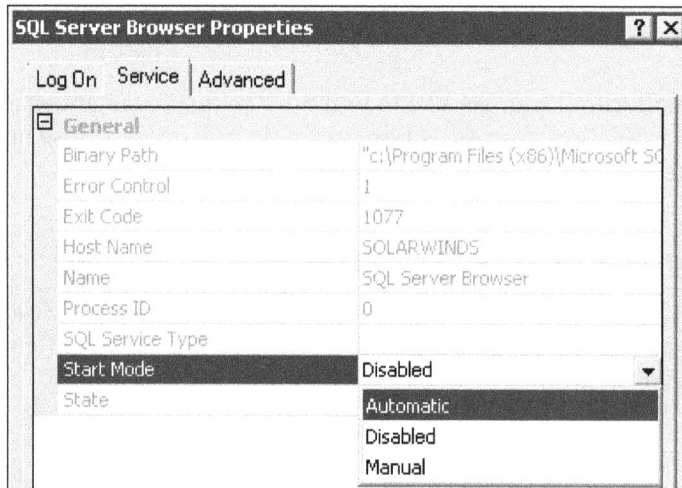

5. Click on **Apply**, then click on the **Log On** tab.

6. Change the option to **This Account** and enter your SQL Server service account information. Click on **Apply**, then on the **Start** button.

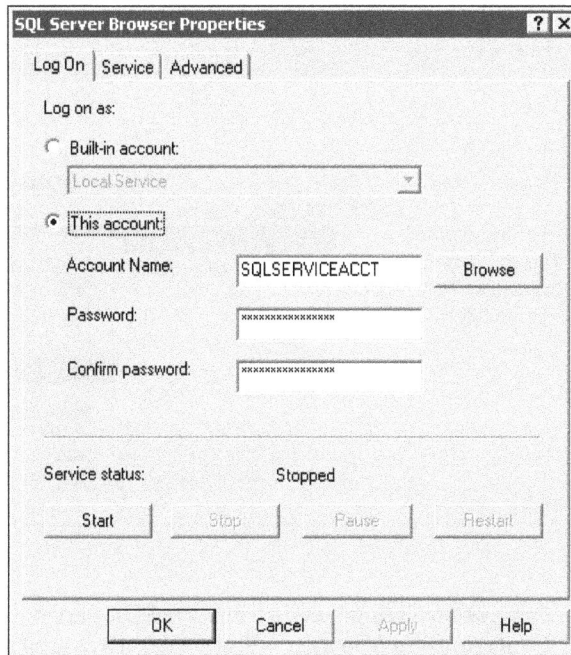

7. Click on **OK** to close **SQL Server Browser Properties**.

8. In the **SQL Server Configuration Manager** window, expand **SQL Server Network Configuration**. Click on **Protocols for <INSTANCE_NAME>**.

9. Verify that **TCP/IP** is **Enabled**. If not, right-click on **TCP/IP** and change the status to **Enabled**.

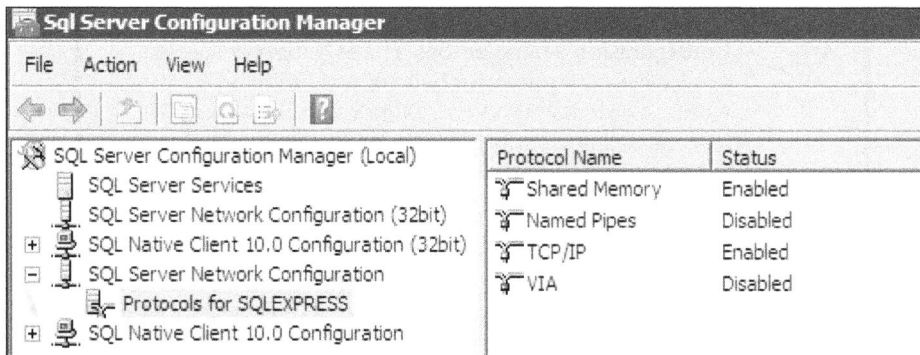

10. In the view pane on the left-hand side, click on **SQL Server Services** to highlight it then right-click on **SQL Server <INSTANCE_NAME>** and choose **Restart**.

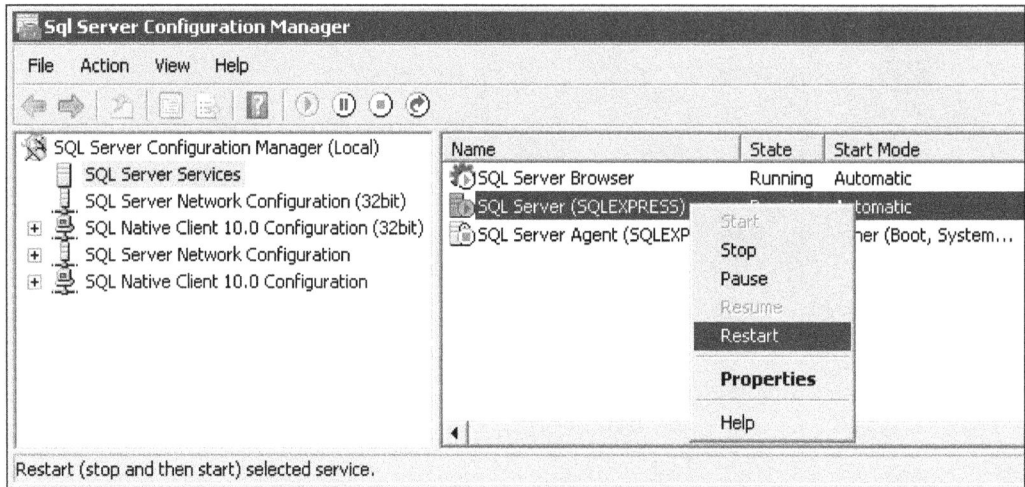

> Any type of change made using the **SQL Server Configuration Manager** will not take effect until the **SQL Server** service has been restarted.

11. Close the **SQL Server Configuration Manager** window.

> The **SQL Server Agent** service is stopped and disabled in this example because the **SQL Server Agent** cannot be enabled in SQL Server Express editions. This is true even though it appears in the **SQL Server Configuration Manager** tool. The **SQL Server Agent** service is not required for SolarWinds Orion NPM to work, so you may ignore it. Do not stop this service if it is enabled and running on Standard, Developer, or Enterprise editions of your SQL Server installation.

Now that the SQL Server services are enabled and configured properly, we can use the SQL Server Management Studio tool to log into the SQL instance and verify that the authentication options are properly configured for SolarWinds Orion NPM.

Verifying SQL Server authentication settings

If you have followed the previous SQL Server installation steps in the previous section, then we have already configured the correct authentication settings. However, it is a good idea to verify that your new SQL Server instance is properly configured even though the SQL installation wizard did all of the work for you. This section will familiarize you in working with the SQL Server Management Studio utility, so I do recommend following the instructions in this section.

1. Log into your SQL Server using the local administrator account of the Windows Server.

2. Launch **SQL Server Management Studio** from the Start Menu.

3. At the login screen, use the following options:

    1. **Server type**: **Database Engine**.

    2. **Server Name**: Use the syntax **Hostname\SQLExpress** where **SQLExpress** is the instance name where you will be creating the Orion NPM database.

    3. **Authentication**: **Windows Authentication**.

    4. Click on **Connect**.

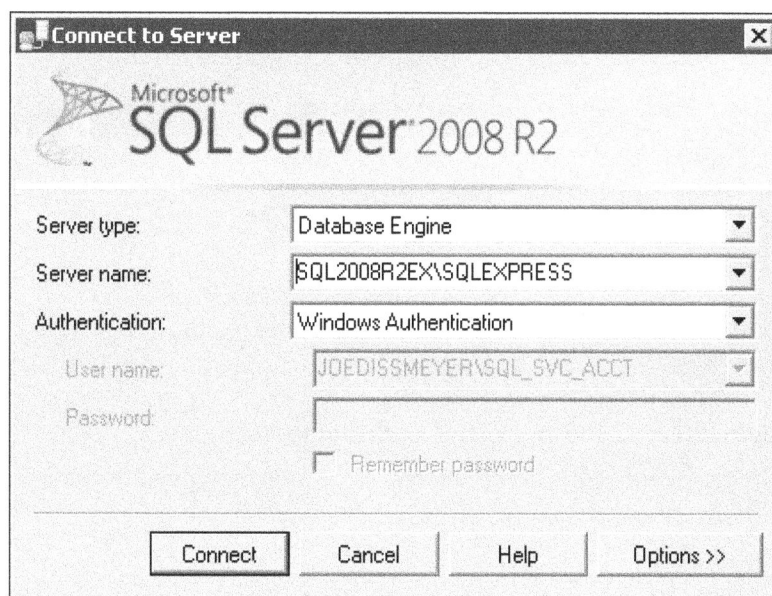

> In this example, I am logged into the Windows Server with the SQL service account. This is the best practice for setting up a SQL Server for the first time. An alternative technique is to log into the Windows Server with its local administrator user account.

4. Under **Object Explorer**, right-click on the SQL Server instance name and choose **Properties**.

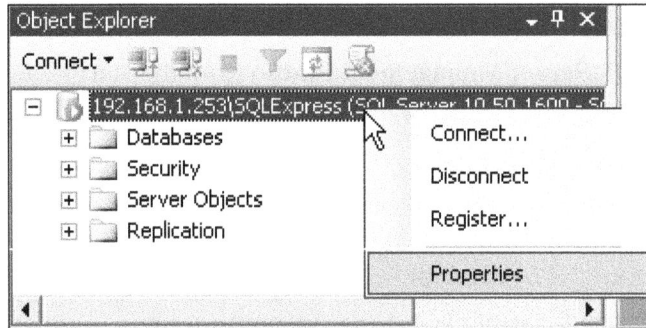

5. The **Server Properties** window appears. Under **Select a page**, click on **Security**. Verify that **SQL Server and Windows Authentication mode** is selected. Click on **Ok** to close the **Server Properties** window.

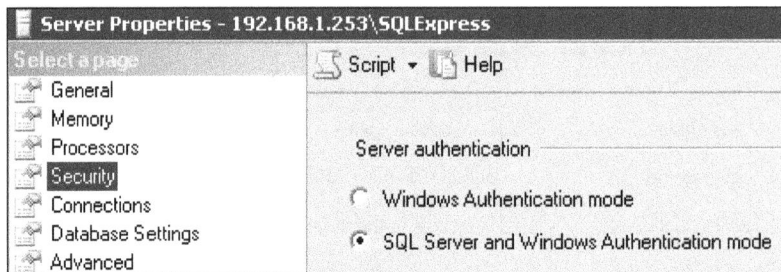

6. Under **Object Explorer**, expand **Security**, then **Logins**.

7. Verify that your SQL Server service account is displayed in the list and has the following permissions:

   ○ **Public**

   ○ **Sysadmin**

   If the SQL service account is not listed, right-click on **Logins** in the **Object Explorer** pane and choose **New Login...** to add it.

8. Close the **SQL Server Management Studio**.

> If you are unsure about the security aspects of SQL Server, Microsoft has excellent documentation and guidance on how to secure SQL Server at `http://msdn.microsoft.com/en-us/library/bb283235.aspx`.

Configuring the Windows Firewall

Now it is time to configure the Windows Firewall on the Windows Server where you will be installing the Orion NPM software as well as on the server running Microsoft SQL Server. If you planned your installation by following SolarWinds' recommendations, you have SQL Server running on one server and Orion NPM installed another.

You must unblock specific TCP/IP ports in the Windows Firewall on each server to allow the SQL Server and Orion NPM software to be able to communicate with each other, as well as receive inbound traffic from nodes when they are polled. If you installed SQL Server and Orion NPM on the same Windows Server, you still must allow specific ports in the Windows Firewall for inbound and outbound polling traffic. This section addresses both of these issues.

If you installed SQL Server on a Windows Server 2008 or 2008 R2 computer, you will use **Windows Firewall with Advanced Security** from Administrative Tools in the Control Panel to allow specific IP traffic into the server. If you installed SQL Server on a Windows Server 2003 or 2003 R2 computer, use **Windows Firewall** from the Control panel.

On the SolarWinds Orion NPM server, allow the following ports for inbound traffic:

*   **TCP 80** and **443** for HTTP, HTTPS, and VMware ESX/ESXi
*   **TCP 17777**, **17778**, and **17779** for Orion NPM Dashboard traffic
*   **UDP 161** and **162** for SNMP
*   **UDP 514** for inbound Syslog messages

> Use **UDP 162** for inbound SNMP traps if you are using a third-party firewall. For the other ports, follow the preceding chart to allow the TCP/IP traffic for both SQL Server and SolarWinds Orion NPM.

Allow inbound traffic to the following programs on the SQL Server only if Orion NPM and SQL Server is running on two different Windows servers:

- `%ProgramFiles% (x86)\Microsoft SQL Server\90\Shared\sqlbrowser.exe`

- `%ProgramFiles% (x86)\ Microsoft SQL Server\MSSQL10.SQLEXPRESS\bin\sqlservr.exe`

> For SQL Server 2008 and 2012 Standard and Datacenter editions, the folder locations will be slightly different within the `%ProgramFiles%\Microsoft SQL Server\` folder. For detailed information, refer to `http://technet.microsoft.com/en-us/library/cc646023.aspx`.

Installing SolarWinds Orion NPM

We are now ready to install SolarWinds Orion NPM. The installation of the Orion NPM software is very straightforward and is very simple. Configuring Orion NPM comes after the installation, which is covered in the next chapter.

Before continuing, ensure that the Microsoft IIS role has not been enabled on the server where you will be installing Orion NPM. We want the Orion NPM installer to set up and configure IIS on our behalf which verifies that the IIS settings are properly configured "out of the box" for Orion NPM.

The installation process for Orion NPM is exactly the same for both Windows Server 2003 and Windows Server 2008 computers regardless of whether they are joined to a domain or not. To get started, log into the Windows Server you will be installing Orion NPM and launch the installer.

1. If you see the following screenshot, click on the **Install** button to continue.

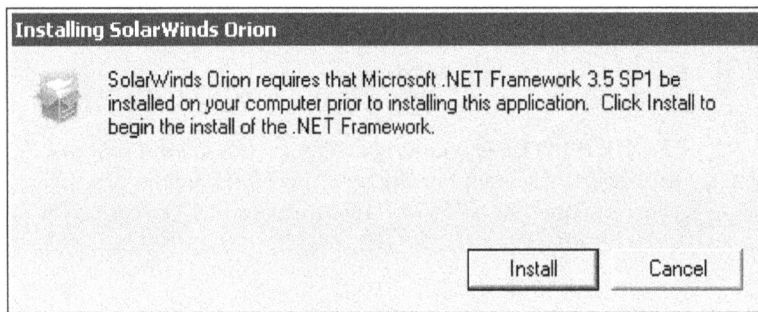

Installing SolarWinds Orion

SolarWinds Orion requires that Microsoft .NET Framework 3.5 SP1 be installed on your computer prior to installing this application. Click Install to begin the install of the .NET Framework.

Install   Cancel

2. The **Help SolarWinds Improve** window appears. Choose yes or no to continue.

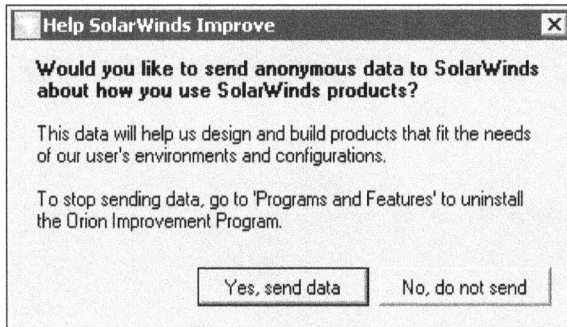

3. On the **Welcome** screen, click on **Next**.

4. Choose which language you want Orion NPM to use, click on **Next**.

5. The **Internet Information Services** screen appears. Recall that earlier in this chapter I stated to make sure that IIS is not installed or enabled on the server. If IIS is disabled, choose **Continue with Orion Installation (Recommended)** and click on **Next** to continue. If IIS is enabled, you will need to stop it, close the installation wizard, remove the IIS role from Windows Server, then restart the Orion NPM installer.

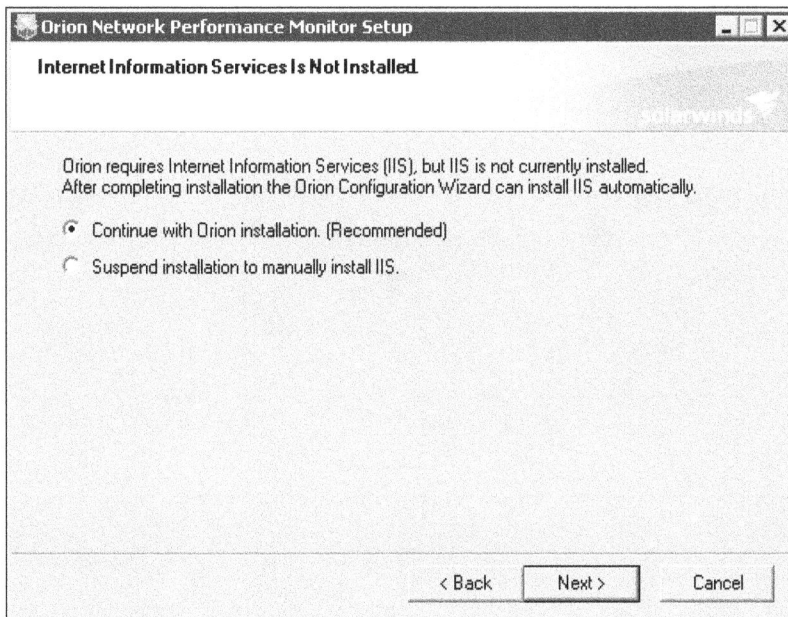

6. Accept the license agreement and click on **Next**.

7. On the **Choose Destination Location** screen, leave the path set to the default **C:\Program Files (x86)\SolarWinds\Orion** and click on **Next**.

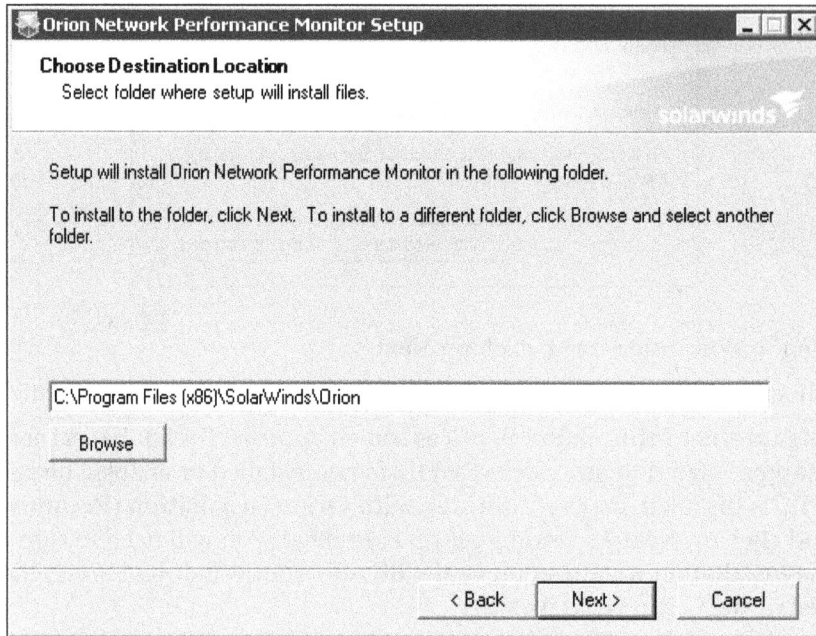

> For 32-bit Windows Servers, the default option will display **C:\Program Files\SolarWinds\Orion**.

8. At the **Start Copying Files** screen, click on **Next** to start the installation.

9. When the Orion NPM installation has completed, click on **Finish** to close the wizard.

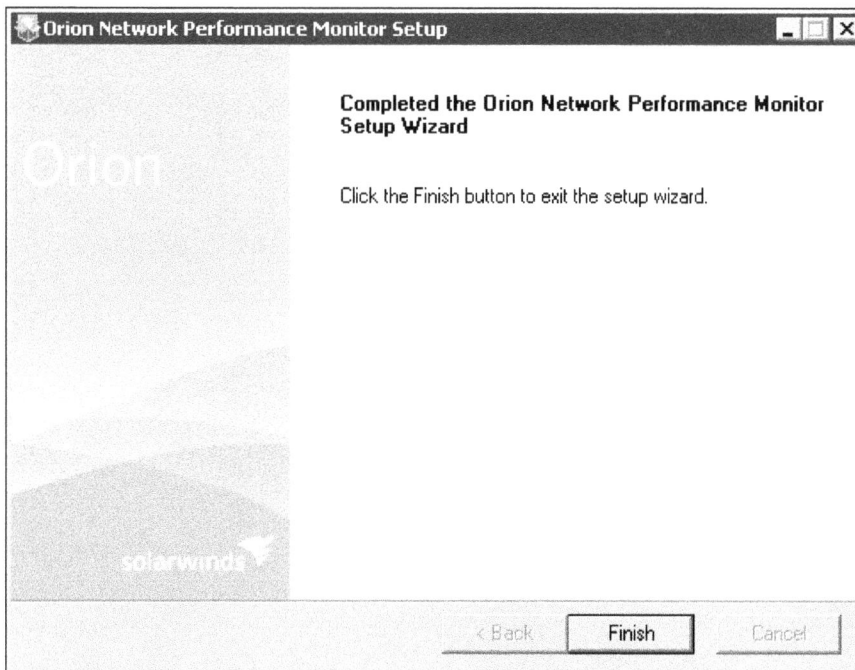

Summary

Congratulations! SolarWinds Orion Network Performance Monitor is now installed and ready for initial configuration. In this chapter we talked about the hardware and server requirements, we walked through an entire SQL Server 2008 R2 Express installation from scratch, configured the Windows Server operating system, configured Windows Firewall settings, and we walked through the SolarWinds Orion NPM installation process.

In the next chapter, we will address Orion NPM's first-time setup using the Setup Wizard.

2

Orion NPM Configuration

In the previous chapter, we walked through an entire Orion NPM installation including the configuration of your Windows Server, the installation and configuration of Microsoft SQL Server, and the Windows Firewall settings. This chapter builds upon the previous one and will cover the initial Orion NPM configuration for a first-time installation scenario. In it, we will discuss the following:

- Orion Configuration Wizard
- Orion Website Administration
- User account permissions
- Setting thresholds
- Orion NPM licensing
- Upgrading Orion NPM from an evaluation license
- An overview of the Orion dashboard

Orion Configuration Wizard

The **Orion Configuration Wizard** handles configuring every single aspect of the Orion NPM product, including the Windows Server itself. Since our Windows Server and SQL Server instance is in order, it is time to launch the Orion Configuration Wizard.

After launching the Orion NPM Configuration Wizard for the first time, a licensing screen appears asking you to click on one of three different buttons. I usually recommend holding off on activating the software until the last possible moment. I do this just in case there is a problem with an installation, if I need to refund a purchase, or if I need to purchase more licensing. But above all, I wait so that I can verify that all settings and configurations are in order.

Perform the following steps while logged into the Windows Server with the same SQL Server service account described when installing SQL server in the previous chapter. Ensure the service account has administrative permissions to the Windows Server where you will be installing Orion NPM. Administrative permissions to this server can be removed when you are finished with the Orion Configuration Wizard.

1. To launch the Orion Configuration Wizard, navigate to **Start | All Programs | SolarWinds Orion | Configuration and Auto Discovery | Configuration Wizard**.

2. The **Licensing Information** window will appear. Click on **Continue Evaluation** to continue. We will activate the software later in *Chapter 8, Maintenance*.

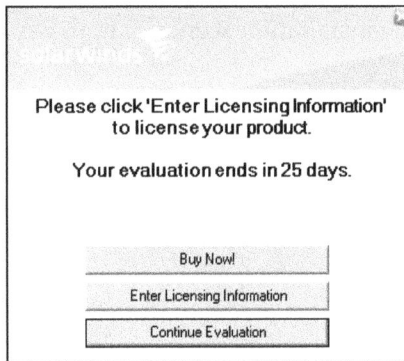

Please click 'Enter Licensing Information' to license your product.

Your evaluation ends in 25 days.

Buy Now!

Enter Licensing Information

Continue Evaluation

3. The Configuration Wizard will display a warning when IIS is not installed. This is the normal behavior. Click on **Yes** to continue.

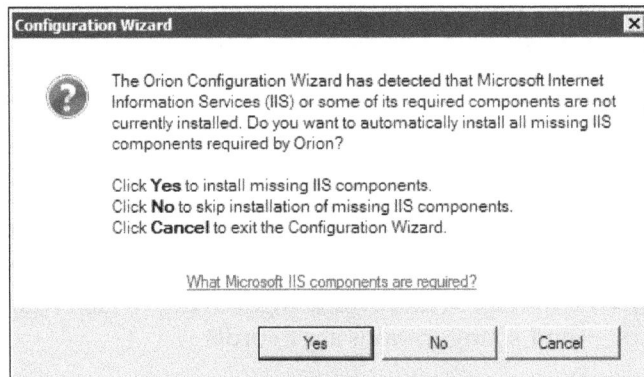

Configuration Wizard

The Orion Configuration Wizard has detected that Microsoft Internet Information Services (IIS) or some of its required components are not currently installed. Do you want to automatically install all missing IIS components required by Orion?

Click **Yes** to install missing IIS components.
Click **No** to skip installation of missing IIS components.
Click **Cancel** to exit the Configuration Wizard.

What Microsoft IIS components are required?

Yes   No   Cancel

> It is at this point where the Orion Configuration Wizard will automatically enable and configure Microsoft IIS. The wizard will continue once the Microsoft IIS role is set up.

4. After a few moments, the **Welcome to the SolarWinds Configuration Wizard** screen will appear. Click on **Next** to continue.

5. At the **Database Settings** screen, choose **Use Windows Authentication**. Click on **Next** when ready to proceed.

> SQL Server has its own local user account database. If you have a local SQL user account with *Public* and *SysAdmin* rights, you may use the **SQL Server Authentication** option instead.

6. Choose **Create a new database** then click on **Next**. It is suggested to use the default database name **SolarWindsOrion** for new installations.

7. At the **Database Account** screen, you must define a local SQL Server account that will bind to the SolarWinds Orion NPM installation. This will be a new user account with rights only to the new SolarWinds Orion database. The account will be created in the local SQL user login account database. Since this is a new installation, choose **Create a new account**, enter a name and a password for the account, then click on **Next** to continue.

Orion NPM does not support domain user accounts for SQL Server authentication. You must choose to create a new account, or choose from an existing one that is local to the SQL login account database. For more information on how to manually create a login account for your SQL Server, see the Microsoft TechNet article at `http://msdn.microsoft.com/en-us/library/aa337562.aspx`.

8. Choose the following Orion website settings.

    1. **IP Address**: Using **All Unassigned** will make the Orion NPM website the default site in IIS. If you have multiple IP addresses assigned to this server, choose one that will be hosting the Orion website.

    2. **Port**: It is recommended to leave it with the default port **80**.

    3. **Website Root Directory**: It is recommended to leave this with the default setting.

    4. **Windows Authentication**: Choose **Yes** to enable logging in using a Windows user account (local or domain), or **No** to only allow login accounts that are local to Orion.

```
SolarWinds Configuration Wizard                                    _ □ ×

Website Settings
    Specify the IP Address, port and location for the Orion website.


    IP Address:              [All Unassigned]          ▼

    Port:                    80

    Website Root Directory:  C:\InetPub\SolarWinds              Browse ...


    Do you want to enable automatic login using Windows Authentication for the Orion Web Console?
    Note: Manual login using Windows Authentication is available in both cases.

    ⊙ Yes - Enable automatic login using Windows Authentication
    ○ No




                                            < Back      Next >      Cancel
```

9. When prompted to create the website root directory, click on **Yes**.

10. At the **Service Settings** screen, ensure that all options are checked and click on **Next**.

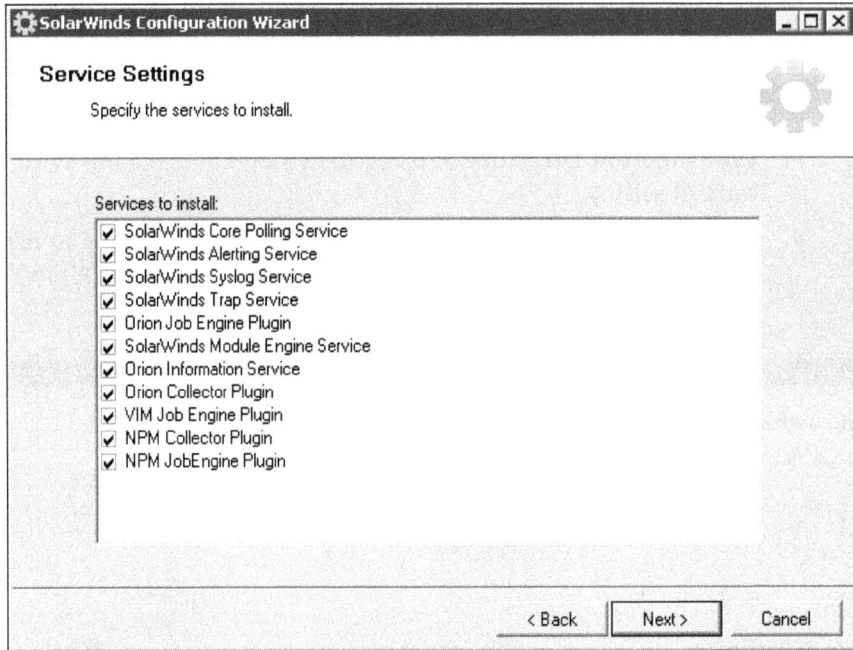

11. Click on **Yes** to disable the Windows SNMP Trap services and enable the SolarWinds Trap Service.

SolarWinds Orion NPM uses its own proprietary service to capture SNMP Traps from monitored devices. This step simply disables the built-in Windows SNMP Trap service and installs the Orion NPM Trap service.

12. Click on **Back** if you need to make any changes. Otherwise, click on **Next** to finish the configuration.

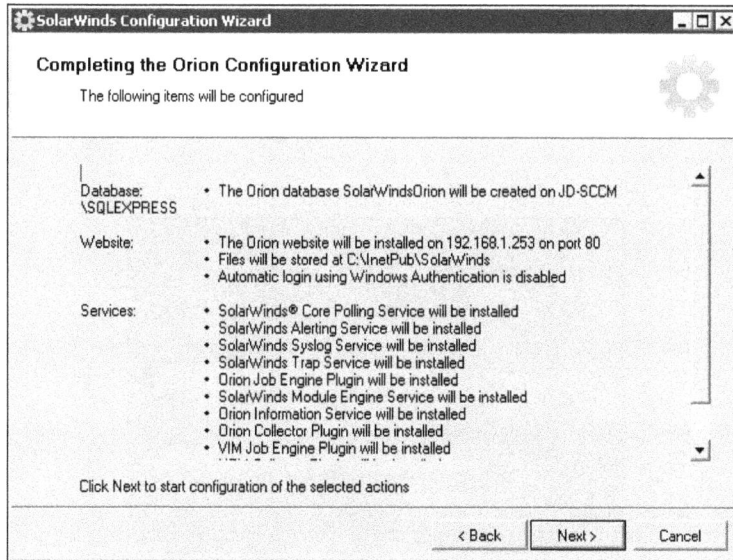

13. The wizard will take several minutes to complete. Click on **Finish** to close the Orion Configuration Wizard.

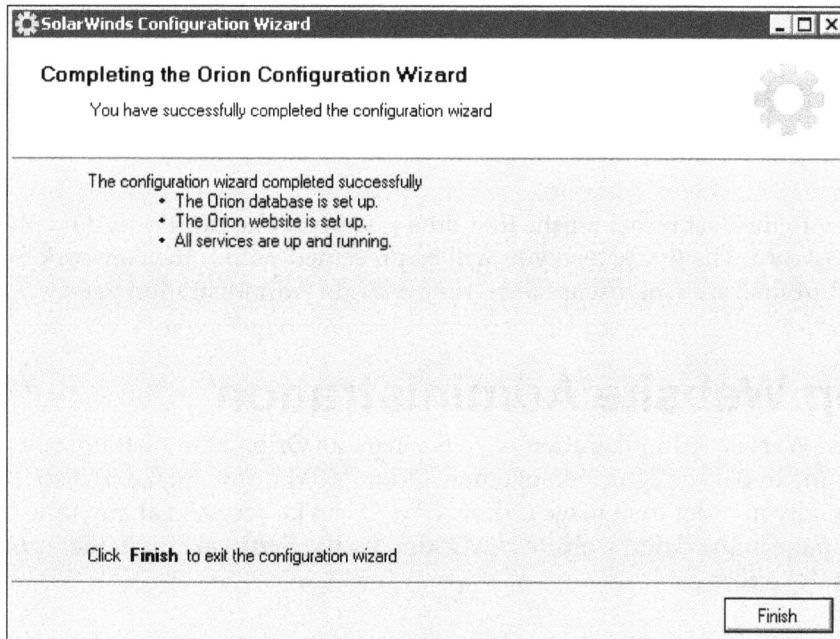

Congratulations! If everything went smoothly, Internet Explorer will automatically open the login page of the Orion dashboard when the Configuration Wizard closes. Remember, you can always come back at a later time and launch the Orion Configuration Wizard if you want to enable or disable any features of Orion NPM.

If you encounter any issues when trying to access the dashboard (such as the dreaded Unable to Display Webpage error), try typing in the IP address and port number you defined in step 7.

Logging into the dashboard for the first time is simple. Use the `admin` ID with a blank password. The first screen you will be presented with is the **Network Sonar Wizard**. But first, we will discuss the Orion Website Administration page.

Orion Website Administration

The Orion Website Administration page is where an Orion administrator not only *completes* the initial configuration of a new Orion NPM installation, but also *returns* to continually in order to manage Orion NPM. It can be accessed at any time from any web page in the Orion website by clicking on the **Settings** link in the upper-right corner of the screen.

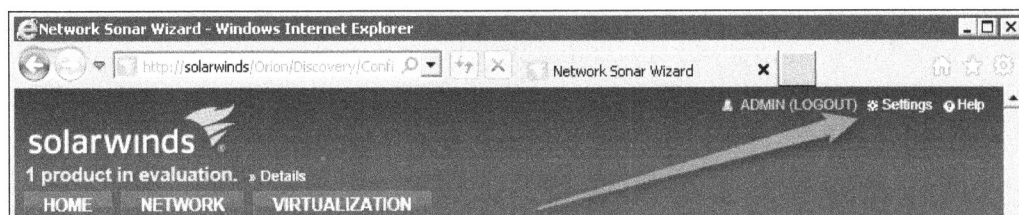

All of the core configuration options for the entire Orion NPM installation are located in the Orion Website Administration console. The admin console allows you to perform the following actions:

- Add nodes and interfaces to Orion NPM either manually or by network discovery
- Manage nodes, groups, and dependencies
- Edit web page views and module views
- Manage Microsoft and VMware credentials
- Configure Orion dashboard settings, polling settings, and thresholds
- Manage user account permissions
- Manage Windows credentials for monitoring Windows-based computers
- Edit the look and feel of the tabs, menu bars, web pages, and modules in the dashboard
- Manage alerts and thresholds
- Check for product updates and view the Orion product team blog
- View details about Orion licensing, the Orion database, and polling engines

Authentication and access

SolarWinds Orion has its own robust user authentication system. Managing user accounts is a routine procedure for an Orion administrator and is very easy to understand when viewing the **Manage Accounts** screen.

After configuring Orion NPM for the first time, you should perform the following two tasks:

- Change the default Admin account password
- Disable the Guest account

Changing the default Admin password is critical in order to secure your new Orion NPM installation. The second task of disabling the Guest account is not required as that account only has read-only access to the Orion website. However, it is wise not to allow access to Orion NPM until you are actually monitoring your network.

Changing the default Admin password

A brand new installation of SolarWinds Orion NPM sets the local Admin account password completely blank. One of the first steps you should take as an Orion administrator is to create a password for this user account. Perform the following steps to change the default administrator password:

1. Open the Orion Website Administration page and under the **Accounts** section click on **Manage Accounts**.

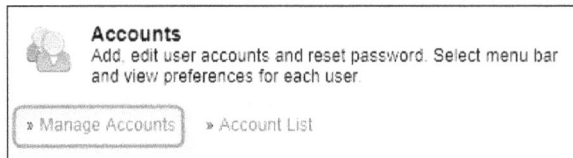

2. On the **Individual Accounts** tab, check the box next to **Admin** and click on the **Change Password** button.

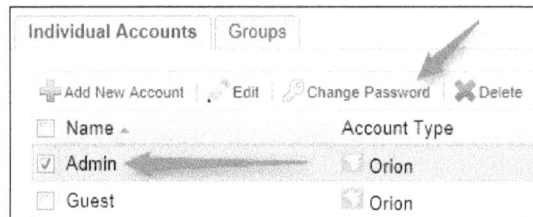

3. Enter a new password and click on the **Change Password** button.
4. Click on **Continue** to finish changing the **Admin** password.

Disabling the Guest user account

Perform the following steps to disable the Guest user account in Orion NPM:

1. On the **Individual Accounts** tab, place a check next to **Guest** and click on **Edit**.

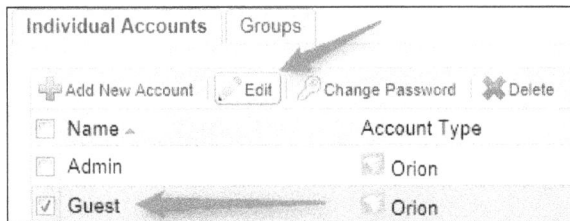

2. Change the **Account Enabled** option to **No**.

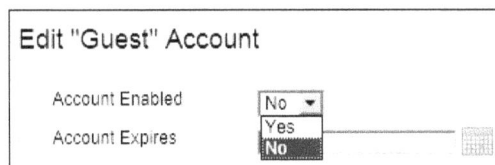

3. Scroll to the bottom of the window and click on the **Submit** button.

The default admin password is now changed and the Guest account is disabled, which will prevent people from accessing Orion NPM.

Orion NPM supports individual user accounts that are local to Orion NPM itself, as well as Active Directory domain accounts. Local Windows accounts, such as a Windows account that is local to the server that Orion NPM is installed on, are not supported.

Individual user accounts are local to the SolarWinds Orion installation and are only specific to Orion NPM. If you try to use a local Windows Server user account to log into the dashboard, it will not work! SolarWinds Orion only allows individual user accounts to be added one at a time, and cannot be added in bulk. If you do not have the need to manage a large number of user accounts in Orion NPM, setting up individual user accounts may be more than sufficient.

An alternative method of setting up user authentication is through Active Directory. **Active Directory (AD)** is a centralized authentication service for Microsoft-based networks and simplifies user account management for the Orion dashboard. Active Directory is a staple in the enterprise and is wildly popular. Enabling Active Directory Authentication in Orion allows you to add both, single domain accounts as well as Active Directory security groups.

The best practice when using Active Directory Authentication is to create a new security group in Active Directory and add domain accounts to the group whom you wish to give access to Orion NPM. Finally, add the security group to the authentication settings in the Orion dashboard. As long as a domain user account is a member of the security group, that user account will, or will not, have rights to the Orion dashboard.

One final thing to know about security groups is that if a user has an individual Orion account and the user is a member of an Active Directory security group added to Orion, the individual Orion user account "wins" and the security group permissions will be completely ignored.

As an Orion administrator, it is up to you to decide how you want to handle user authentication to your Orion NPM system. You can always start with assigning individual accounts, then implement Active Directory Authentication at a later date. Or, you can use both methods at the same time. There is no "user licensing" for SolarWinds Orion NPM. You can create/add/edit as many user accounts as you want!

> If you have the means of setting up Active Directory account management, it is highly recommended to do so as it eases the administrative burden of user access to Orion NPM. The Active Directory Authentication feature is enabled through the Orion Configuration Wizard.

Manage Accounts overview

You can manage user account access in two ways, by using individual accounts or Active Directory security groups.

Name ▲	Account Type	Enabled	Expiration	Last Login	Account Limitation	Admin Rights	Node Mgmt	View Customization
Admin	Orion	Yes	Never	Wednesday...	None	Yes	Yes	Yes
Guest	Orion	Yes	Never	Never	None	No	No	No
JOEDISSMEYER\JOEDISSMEYER	Windows	Yes	Never	Never	None	Yes	Yes	Yes
JOEDISSMEYER\LAURENJONES	Windows	Yes	Never	Never	Single Network Node	No	Yes	Yes
JOEDISSMEYER\MarkJohnson	Windows	Yes	Monday, December 31, 2012...	Never	None	No	No	Yes
JohnAdams	Orion	Yes	Never	Never	None	Yes	No	No

The **Individual Accounts** tab shows how easy it is for an Orion administrator to see who has been logging into the dashboard, what type of user accounts have access to the dashboard, what user accounts may need to be deleted due to inactivity, and a general idea of what permissions a user account may have.

Active Directory security groups are added to Orion from the **Groups** tab in the **Manage Accounts** page. The process for adding a group, as well as assigning permissions to a group, is precisely the same as individual accounts. Groups can be added only if Active Directory authentication was enabled from the Orion Configuration Wizard.

	Orc Name	Enabled	Expiration	Last Login	Account Limitation	Admin Rights	Node Mgmt	View Customization
1	JOEDISSMEYER\Orion NPM Users	Yes	Never	Never	None	No	No	No
2	JOEDISSMEYER\Orion NPM Node Administrators	Yes	Never	Never	Group of Nodes	No	Yes	Yes
3	JOEDISSMEYER\Orion NPM Virtualization Administrators	Yes	Never	Never	None	No	Yes	Yes
4	JOEDISSMEYER\Orion System Administrators	Yes	Never	Never	None	Yes	Yes	Yes

Creating individual user accounts

Use the following example to create an individual user account:

1. From the Orion dashboard, click on the **Settings** link on the upper-right corner of the window.

2. Under the **Accounts** module, click on **Manage Accounts**.

3. Select the **Individual Accounts** tab and click on the **Add New Account** button.

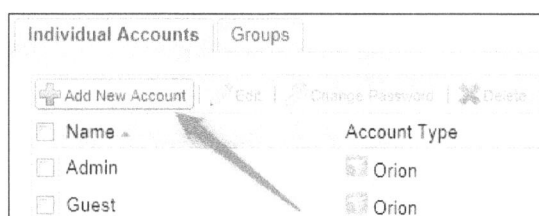

Name ▲	Account Type
Admin	Orion
Guest	Orion

4. Choose the **Orion individual account** option and click on **Next**.

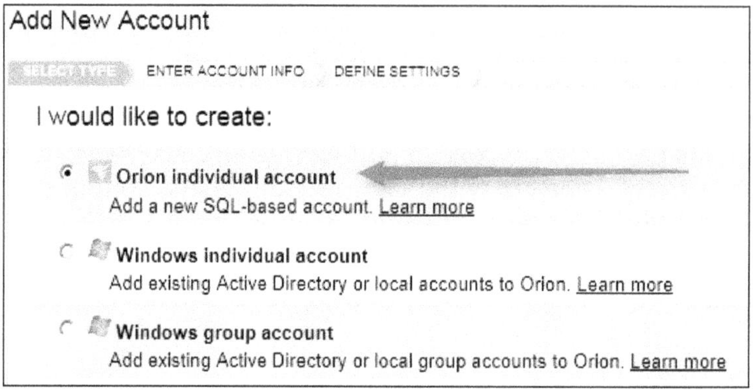

Add New Account

SELECT TYPE   ENTER ACCOUNT INFO   DEFINE SETTINGS

I would like to create:

○ Orion individual account
   Add a new SQL-based account. Learn more

○ Windows individual account
   Add existing Active Directory or local accounts to Orion. Learn more

○ Windows group account
   Add existing Active Directory or local group accounts to Orion. Learn more

5. Type in a user name and password and click on **Next**.

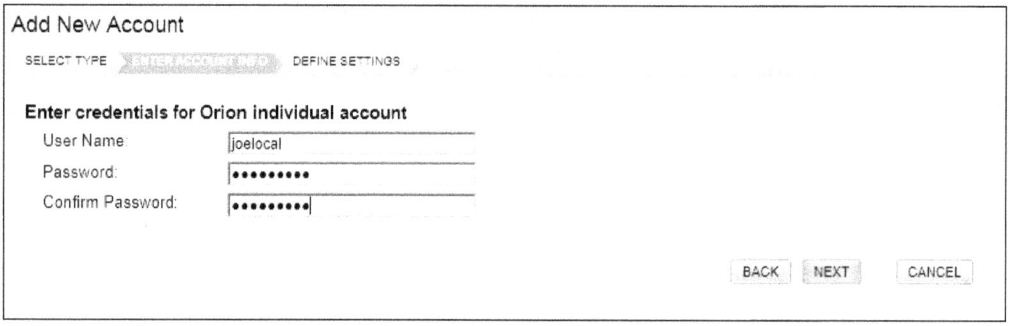

Add New Account

SELECT TYPE   ENTER ACCOUNT INFO   DEFINE SETTINGS

Enter credentials for Orion individual account

User Name: joelocal
Password: •••••••••
Confirm Password: •••••••••

BACK   NEXT   CANCEL

6. Define the permissions for this new user account. When finished, scroll to the bottom of the page and click on **Submit**.

Add New Account

SELECT TYPE   ENTER ACCOUNT INFO   **DEFINE SETTINGS**

Define settings for Orion individual "joelocal" account

Account Enabled	Yes ▾	Disabled accounts cannot log in.
Account Expires	Never	This account cannot log in after this date. Enter "Never" for accounts that should not expire.
Disable Session Timeout	No ▾	If session timeout is disabled, this account will stay logged in indefinitely even if the browser is closed. You can still click logout to end your session securely.
Allow Administrator Rights	No ▾	Accounts with Admin rights can Add and Edit other Accounts and reset passwords.
Allow Node Management Rights	No ▾	Accounts with Node Management role can manage nodes.
Allow Account to Customize Views	No ▾	Enable this to allow the Account to customize the Views. Any changes made to a View are seen by all Accounts with the same View.
Allow Account to Clear Events, Acknowledge Alerts and Syslogs	Yes ▾	Enable this to allow the Account to Acknowledge/Clear Events from the Event Log, as well Acknowledge Advanced Alerts from the Alerts view and Syslogs Messages from the Syslog view.
Allow Browser Integration	No ▾	Browser Integration allows you to utilize tools on the client browser machine with information provided by the web page. Links in most web pages will then be right-clickable, offering you a choice of options to perform on the device that the link represents. Community strings will not be sent to the browser unless "Allow Secure Data on Website" (in the System Manager) is enabled.
Alert Sound	No Alert Sounds ▾	Set this to a valid .wav file to enable audible alerts via the web browser.
Number of items in the breadcrumb list	50	If this value is set to 0 - all items will be shown within breadcrumbs drop down list.

Account Limitations

There are no account limitations defined. To create account limitations, click the "Add Limitation" button.

The new user account will be listed in the **Individual Accounts** tab. If you make a mistake, you can always go back and edit the account permissions. Otherwise, delete the account then create a new one.

Adding Active Directory user accounts

Use the following example to grant an Active Directory domain account access to the Orion dashboard:

1. Select the **Individual Accounts** tab and click on the **Add New Account** button.

2. Choose the **Windows individual account** option and click on **Next**.

3. Under the **ACTIVE DIRECTORY OR LOCAL DOMAIN AUTHENTICATION** section, enter the credentials of a domain user account that has administrative access to the Active Directory database. The following example shows me my user ID JOEDISSMEYER from the JOEDISSMEYER.LOCAL domain.

You must type in the domain name first when defining Active Directory account information in Orion NPM. Also, you must have a domain user account that has rights to read user account information from Active Directory. At a minimum, the account needs read-only rights to the domain.

4. Scroll down to the **SEARCH FOR ACCOUNT** section, enter the domain user account in the textbox using the DOMAIN\USERNAME syntax, then click on the **Search** button.

SEARCH FOR ACCOUNT		
User name:	joedissmeyer.local\laurenjones	Search
	Use Domain Username format	

5. Under the **ADD USERS** section, place a check mark next to the user ID to add it to the list on the right-hand side. Click on **Next** when you are ready to continue.

ADD USERS		
Select users to add:	☐ Account Name ▴	▴ 3 accounts selected
	☑ JOEDISSMEYER\Administrator	JOEDISSMEYER\Administrator ✖
	☑ JOEDISSMEYER\JOEDISSMEYER	JOEDISSMEYER\JOEDISSME... ✖
	☑ JOEDISSMEYER\laurenjones	JOEDISSMEYER\laurenjones ✖
	☐ JOEDISSMEYER\SQL_SVC_ACCT	
	4 matches found	

You can add as many Active Directory user accounts as you like at once. Simply clear the **SEARCH FOR ACCOUNT** box and type in a new user ID to search for, and then check the box to add the user ID to the list to the right-hand side.

If your search for an account fails, try using the FQDN of the domain instead (that is, DOMAIN.COM instead of just DOMAIN). You can also use "fuzzy" search terms that include the first or last name of the account. For example, you can search for joedissmeyer\joe instead of the full user ID.

6. Define the permissions for this new user account then click on **Next**. Remember that if you added multiple accounts, this will define permissions for all of them.

The new user account(s) will be listed in the **Individual Accounts** tab. Notice that the **Account Type** states it is a **Windows** domain user account. Accounts that are local to the Orion server will say **Orion**.

Name ~	Account Type
Admin	Orion
Guest	Orion
JOEDISSMEYER\LAURENJONES	Windows
JOEDISSMEYER\MarkJohnson	Windows
MickeyMouse	Orion

If you made a mistake when adding the accounts in the dashboard, you can always go back and edit or delete them.

Adding Active Directory security groups

Adding Active Directory groups to Orion can be done in one of the two ways; directly from the **Groups** tab or from the **Individual Accounts** tab. But, for the sake of simplicity, we will see how to do this from the **Groups** tab.

1. Click on the **Groups** tab, then click on the **Add New Group Account** button in the menu bar.

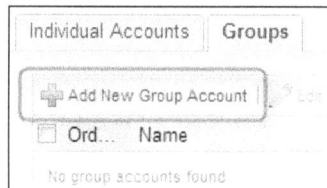

Individual Accounts	Groups
Add New Group Account	
Ord... Name	
No group accounts found	

2. Under the **ACTIVE DIRECTORY OR LOCAL DOMAIN AUTHENTICATION** section, enter the credentials of a domain user account that has administrative access to Active Directory.

3. Under the **SEARCH FOR ACCOUNT** section, enter the domain group account in the textbox using the `DOMAIN\GROUPNAME` syntax and then click on the **Search** button. If your search fails, try using the FQDN of the domain instead of its common name.

SEARCH FOR ACCOUNT		
Group name:	joedissmeyer.local\orion	Search
	Use Domain\Groupname format	

4. Under the **ADD GROUPS** section, place a check mark next to the group name to select it. If you want to add more groups at this time, clear the **SEARCH FOR ACCOUNT** textbox and type in a new name to search. Click on **Next** to continue.

ADD GROUPS

Select groups to add	Account Name ▲	▲ 2 groups selected
	☐ Account Name ▲	JOEDISSMEYER\Orion System... ✖
	☑ JOEDISSMEYER\Orion NPM Node Administrators	JOEDISSMEYER\Orion NPM N... ✖
	☐ JOEDISSMEYER\Orion NPM Users	
	☐ JOEDISSMEYER\Orion NPM Virtualization Admini...	
	☑ JOEDISSMEYER\Orion System Administrators	

4 matches found

5. Define the permissions which will be assigned to this addition then click **Next**. Remember that if you added multiple groups, Orion will define permissions for all that were added at this time.

Your Active Directory groups will be listed in the **Groups** tab. Again, remember that if a user account has an individual account in Orion and is also a part of the Active Directory security group, Orion will ignore the group permissions.

Account permissions

SolarWinds Orion is fully compliant with the concept of "the principle of least privilege". It allows an administrator to be extremely granular regarding permissions to tabs, pages, and modules. Almost every aspect of the Orion dashboard has a user permission which can be allowed or denied per user account or per Active Directory group. It is even possible to limit one user account to a single node! You would want to limit access because some modules actually allow you to change a setting on the actual node. You wouldn't want a virtualization administrator to start making changes to your core production routers, would you? If you need to know what options allow or deny, a full description for each permission is displayed in the account edit screen.

Setting thresholds

Prior to adding nodes to Orion NPM, thresholds need to be configured. At this point you may be asking, "Joe, what is a threshold?" A **threshold** is the level at which Orion will take action on a given value. For example, if one of your monitored servers has a CPU load of more than 80 percent, Orion will warn you about the problem in the dashboard. After the CPU load is more than 90 percent, Orion will trigger an alert action, such as sending a text message to a mobile device or an e-mail notification. Thresholds are how Orion knows how to warn and notify against certain situations. It is up to you to make sure that the threshold levels are not set too high so that events are not missed, or set too low so that Orion doesn't generate a great deal of alerts under what may be considered "normal" conditions. The default threshold settings in a new Orion NPM installation are more than suitable for almost every situation. However, it is still a good idea to look over the levels in case you feel something needs to be changed in the future.

There are two levels of thresholds, the **warning** and **high** level.

- The warning level notifies an administrator in the dashboard allowing you to take a "pre-emptive strike" action to resolve the issue on a node before the situation has a chance to get out of hand.

- When a node reaches the high level, an Orion alert will be generated and triggers an action, such as sending a text message to an administrator's mobile device or sending an e-mail (or both). Almost all threshold settings are set to a percentage.

Beyond the two threshold levels, there are three different threshold types as follows:

- Orion general thresholds
- Network performance monitor thresholds
- Virtualization thresholds

Orion general thresholds are thresholds that encompass every single node that is monitored in Orion. These are the "general" monitored levels for *everything*. Even certain devices that do not necessarily fall into a specific category, such as a battery backup unit, are still bound to these default threshold rules. The following are the five general thresholds:

- Average CPU load
- Disk usage
- Percent memory used
- Percent packet loss
- Response time

Every general threshold is measured by a percentage except for response time, which is measured by milliseconds. The settings for general thresholds are global settings for Orion NPM. You cannot change general thresholds for different nodes or different interfaces. If you are not sure what threshold settings you may need to change, it is suggested to leave the default options. You can always come back and edit thresholds as you see fit.

Admin ▸ Settings ▸			Help

Orion General Thresholds

Configure node and volume thresholds for all Orion modules.

Avg CPU Load

High Level	90	1% to 100%	Nodes with CPU Load above this level will appear on "High CPU Load" reports. The Gauges will also be colored Bold Red.
Warning Level	80	1% to 100%	Nodes with CPU Load above this level will appear on "High CPU Load" reports. The Gauges will also be colored Red.

Disk Usage

High Level	95	1% to 100%	Disk Usage will be colored Bold Red and appear on "High Disk Usage" reports when Percent Disk Usage is above this level.
Warning Level	80	1% to 100%	Disk Usage will be colored Red and appear on "High Disk Usage" reports when Percent Disk Usage is above this level.

Percent Memory Used

Error Level	90	1% to 100%	Nodes with Percent Memory Used above this level will appear on "High Percent Loss" reports. The Percent Loss Gauges will also be colored Bold Red.
Warning Level	80	1% to 100%	Nodes with Percent Memory Used above this level will appear on "High Percent Loss" reports. The Percent Loss Gauges will also be colored Red.

Percent Packet Loss

Error Level	50	1% to 100%	Nodes with Percent Packet Loss above this level will appear on "High Percent Loss" reports. The Percent Loss Gauges will also be colored Red.
Warning Level	30	1% to 100%	Nodes with Percent Packet Loss above this level will appear on "High Percent Loss" reports. The Percent Loss Gauges will also be colored Red.

Response Time

Error Level	1000	1ms to 5000ms	Nodes responding above this level will appear on "High Response Time" reports. The Gauges will also be colored Bold Red.
Warning Level	500	1ms to 5000ms	Nodes responding above this level will appear on "High Response Time" reports. The Gauges will also be colored Red.

The **NPM thresholds** are specific to all nodes in the **NETWORK** tab and apply in addition to the general threshold settings. Nodes that NPM thresholds would apply to include switches, routers, firewalls, servers, and any other network node type. There are three options in the NPM thresholds:

- Cisco buffer misses
- Interface errors and discards
- Interface percent utilization

Again, the default general threshold options will suffice for almost every
network environment.

The final threshold type is **Virtualization Thresholds** for VMware monitored nodes.
There is only one threshold to configure, **VMware Network Utilization**, and it
applies in addition to the general threshold settings for VMware nodes.

Orion NPM licensing

Orion NPM has a flexible licensing model and is licensed by the highest number of the following elements:

- Nodes
- Interfaces
- Volumes

An **interface** element is some type of primary point of contact with the network. Switch ports, Serial Ports, VLANs, Port Channels, and both physical and virtual network interface cards (NICs) are all classified as interfaces.

A **node** is a physical or virtual network device that can be managed. Switches, routers, physical servers, virtual servers, virtual server hosts (VMware, Xen, and so on), firewalls, wireless access points, wireless controllers, and other types of physical network devices are all classified as node elements in Orion NPM. For example, one wireless access point monitored by SNMP would be classified as a node. A Windows Server monitored by WMI would be a node as well. In general, a node element has more than just an interface or volume. A node may have a fan, CPU, memory, power supplies, and other resources. In general, if a network device can be monitored by SNMP, Syslog, or WMI, that device would use a node element license in Orion NPM and all of its resources would be monitored.

A **volume** is a storage medium that a computer uses to store information against some type of filesystem. RAM or memory is a volume, disk partitions are volumes, and the swap partition in Linux is also a volume. Volumes are licensed by each separate logical disk in Orion NPM. If one physical hard disk in one server is split into four volumes, then Orion NPM assigns four volume licenses to that server. Another example could be a **Storage Area Network** (**SAN**) with twenty hard drives split into one hundred volumes. The SAN would consume one hundred volume licenses in Orion NPM.

Monitoring resources beyond interfaces and volumes are not counted in the Orion NPM licensing model. According to SolarWinds' official documentation,

> *Data collected for items other than interfaces, nodes, and volumes is effectively free.*

So if you were wondering about having to pay extra just to monitor and alert against high CPU usage, or a fan failure, you need not worry any longer.

> The official SolarWinds Orion NPM Product Guide can be found at
> `http://www.solarwinds.com/network-performance-monitor/`
> `resources.aspx#ProductGuides.`

License tiers

Orion NPM and is sold in licensing packages, or **tiers**.

License Tier	Description
SL100	Licensed for 100 interfaces, and 100 nodes, and 100 volumes
SL500	Licensed for 500 interfaces, and 500 nodes, and 500 volumes
SL2000	Licensed for 2000 interfaces, and 2000 nodes, and 2000 volumes
SLX	Unlimited amount of interfaces, nodes, and volumes

Selecting the correct license tier is important when planning your Orion NPM implementation. Ensure that you purchase the license tier for the maximum amount of elements you plan on monitoring on your network. For example, if you have 1500 interfaces, 1000 nodes, and 700 volumes, then you would only need the SL2000 license because the highest count of elements is interfaces.

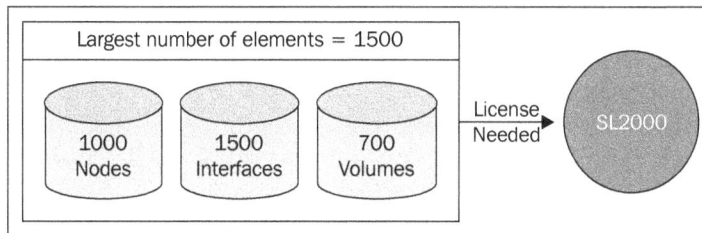

The SLX license is special because you can monitor an unlimited number of elements with this license. However, the SLX licensing limits the throughput of its polling engine since SLX is intended for very large networks. The reason why this throughput is limited is to reduce slowdown within Orion NPM and reduce network latency. You wouldn't want complaints of network performance issues on your network, would you?

Extending and increasing NPM licensing

The SL tier licenses cannot be stacked or extended on the same server. For example, if you have an SL500 license as well as an SL2000 license, you cannot activate both licenses on the same server to increase the total count to 2500 elements. Instead, you must upgrade to the next level license tier, SLX. The only alternative is to install an additional Orion NPM server in your environment. If you have an SL100 license but only need to extend the licensing of 100 more elements, you must purchase the SL500 license to extend Orion NPM monitoring to a total of 500 elements.

The maximum element licensing model is unique to SolarWinds Orion products and, seeing how network devices are built and monitored, this model makes sense. Network devices and servers such as load balancers, firewalls, routers, and even large switches come in a variety of sizes, features, and configurations, with each having multiple uplink ports, modules, and hard drives with multiple volumes. It is possible for a single device to consume multiple Orion NPM element licenses for all three element types.

Upgrading Orion NPM from an evaluation license

Orion NPM can be downloaded and used for 30 days for free without a license key. The 30-day trial is a full install with no limits placed on its features. So if you want to take the software for a trial run before committing to a purchase, feel free to do so. In fact, SolarWinds encourages you to try out the software first before committing to a purchase. If you do plan on keeping Orion NPM, it does not need to be un-installed then re-installed all over again if you decide to buy the product. When you buy a license key during the trial period, simply activate the SL or SLX license key to move it from trial mode to a fully-licensed product.

> Orion NPM can be downloaded for free from
> `http://www.solarwinds.com/downloads/`.

Orion NPM will simply cease monitoring your network after the trial period has expired. The good news is that all of your device configuration, reports, historical data, and anything stored by Orion NPM will not be deleted. But the dashboard will not display any data or reports and you will not be able to add or delete any nodes. The dashboard will only prompt you to activate a license. Once you activate Orion NPM with a valid license key, it will start monitoring your network immediately. However, keep in mind that there might be a gap in monitoring data from when Orion NPM stopped monitoring your network to when Orion NPM was activated.

After you are confident that your Orion NPM configuration is running as expected, activate your license key. Orion NPM cannot be activated from the dashboard. Instead, this task is completed using the **Network Performance Monitor Licensing** utility which is an application on the Windows Server where Orion NPM is installed on. Using the licensing utility, you can activate Orion NPM using an Internet connection or manually via the SolarWinds Customer portal. Your paid license key is also available in the SolarWinds Customer Portal.

Once you have your license key, follow the next set of instructions to upgrade your evaluation to a fully-licensed Orion NPM installation:

1. Retrieve your license key by logging into the SolarWinds Customer Portal with your customer account. The customer portal website is located at `http://www.solarwinds.com/customerportal/`.

2. Log into the Windows Server where Orion NPM is installed.

3. Launch the **Network Performance Monitor Licensing** utility from the Start Menu.

4. Click on the **Enter Licensing Information** button.

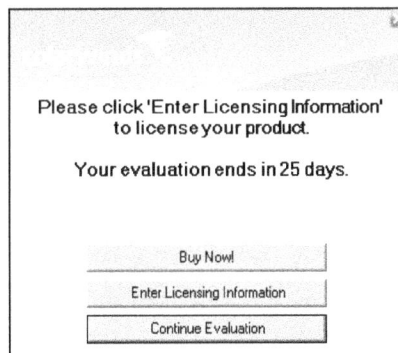

5. Choose your activation method. If your server has an active Internet connection, choose the first option, enter your license key and proxy server information (if applicable), and click on **Next** to activate.

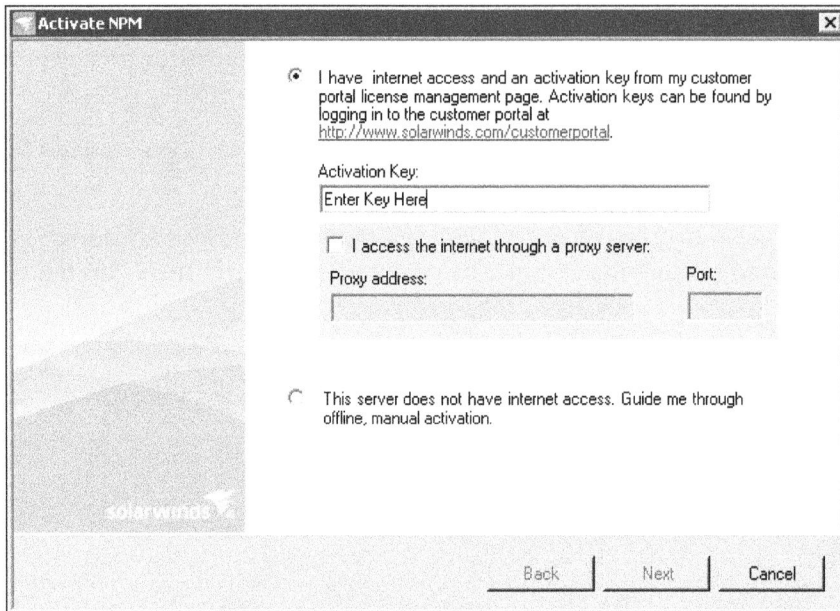

1. If you wish to activate it manually, or if your server does not have an active Internet connection, choose **This server does not have internet access. Guide me through offline, manual activation** and click on **Next**.

2. Follow the instructions on the following screen to activate Orion NPM:

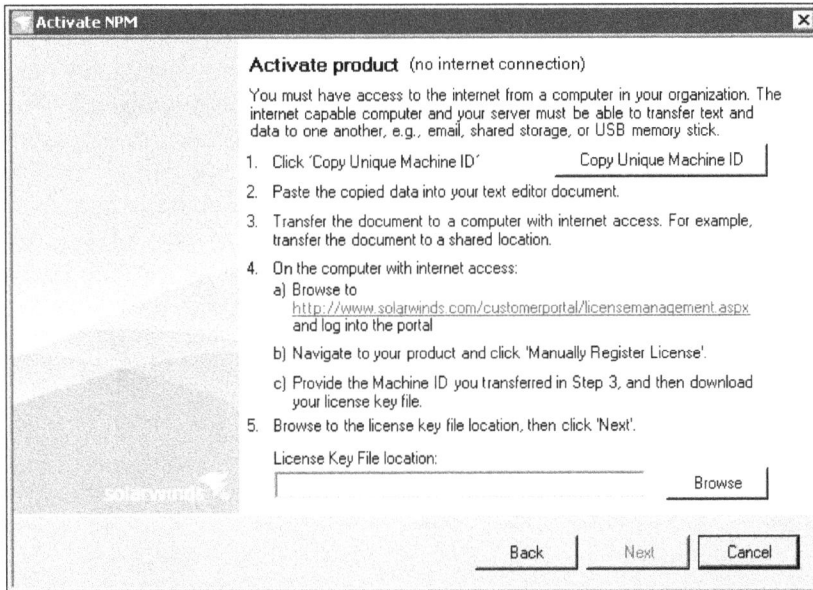

3. Click in the **Copy Unique Machine ID** button to copy the information to your clipboard.

4. Save this ID to a text file on a USB stick.

5. On a computer with Internet access, log into the SolarWinds Customer Portal with your SolarWinds customer account and click on the **License Management** link.

6. Choose your product from the list and click on **Manually Register License**.

7. Using the machine ID you saved from the offline server, register the machine ID and retrieve the license key file from the portal.

8. Transfer the license key file to the offline computer and apply it using the **Activate NPM** window.

You may be prompted to restart the Orion NPM services on the server to finish activation.

Licensing, best practices

When a node is added to Orion NPM, you must pick and choose what elements and resources you want to monitor. It is at this point where you must be mindful of your licensing limits. A best practice for Orion NPM licensing is *monitor only what you need to*. If you choose to monitor all 52 ports of a switch, all VLANs, and Port Channels, you will surrender at least 54 interface licenses including one node license. In most cases, you will only need to monitor one or two trunk ports from an endpoint device and nothing more. If you want to monitor all elements of a server with two hard drives and two network interfaces, then you will have to surrender two interface licenses, one node license, and at least two volume licenses. For most servers, you may only need to monitor the node for up/down availability. Again, only monitor exactly what you *need* so that you don't exceed your licensing.

> You cannot monitor an interface or volume on a node without also monitoring the node itself.

Overview of the dashboard

Now that you are logged in, you can start working with Orion NPM through the web console or the Orion dashboard. The dashboard is where users will perform most of the duties as an Orion administrator. It operates by using a hierarchical tabbed-based interface. There are three primary tabs at the top of the Orion dashboard; **HOME**, **NETWORK**, and **VIRTUALIZATION**. Every one of these primary tabs has a bar that displays different web page links. Every page has their own set of modules that shows the specific aspects of that area. For example, the **SUMMARY** page in the **HOME** tab displays the overview of the monitored network.

> SolarWinds refers to the Orion dashboard as both "the dashboard" and "the web console". These terms are used interchangeably.

One of the best features of Orion NPM is that virtually every single tab, page, and module in the dashboard can be manipulated, edited, changed, or moved from one page to another if an Orion administrator wishes to do so. In addition, user account access can be allowed or prohibited to any tab, page, or module.

As you navigate, the dashboard will display breadcrumbs to lead you back to a previous page, or back to the tab or page where a module resides. For example, from the **SUMMARY** page in the **HOME** tab, I can click on the module titled **Node Management**. The breadcrumb will lead me back to the Orion Website Administration page since that is where the module truly resides.

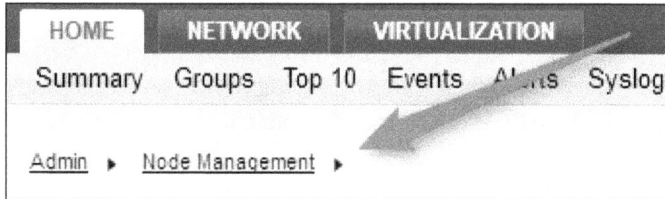

Website notifications

From time to time, a notification bar may appear underneath the menu bar displaying information about product updates, new SolarWinds company blog posts, new node discoveries, licensing information, and other tasks. Depending on the notification displayed, a hyperlink will take you to the corresponding administrative page. The following example shows a notification about a scheduled network discovery job that has recently completed. All of the various notifications that Orion can display in the dashboard can be disabled via Orion Website Administration.

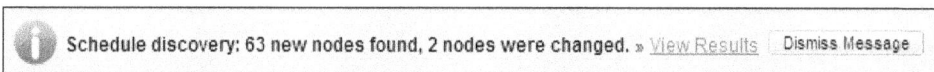

The following is another example of a notification about SolarWinds Orion Team's blog posts, as well as new available product update:

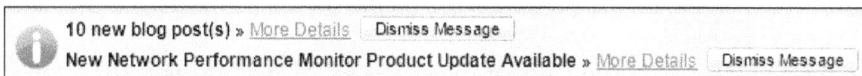

HOME tab

The **HOME** tab and each page in its link bar are items that provide information on the Orion NPM system as a whole. Many of the pages in the **HOME** tab are suitable for IT management and help desk personnel to help give them the "big picture" about the current status of your network.

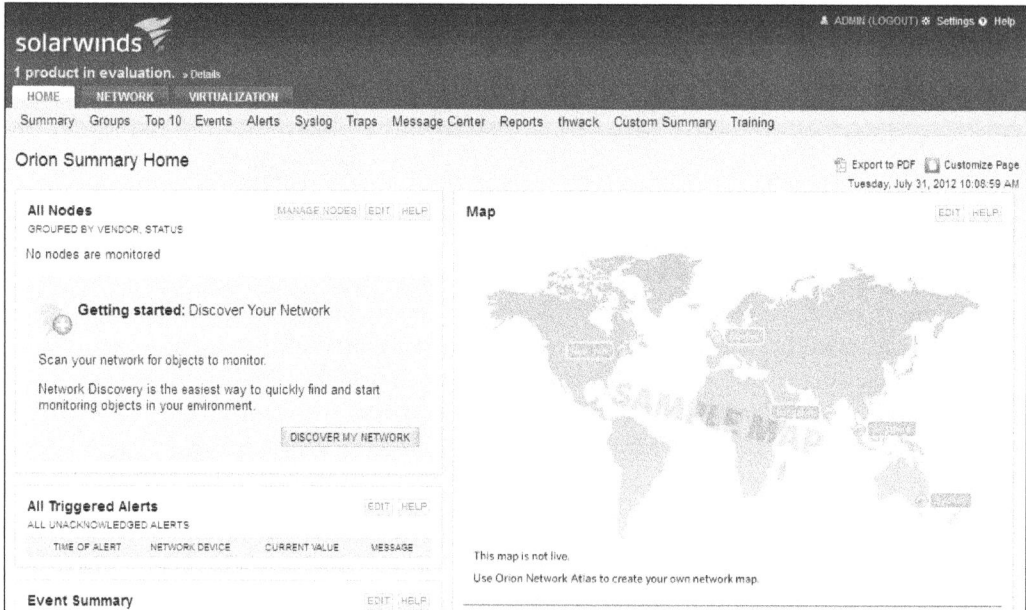

The link bar pages in the **HOME** tab are **Summary**, **Groups**, **Top 10**, **Events**, **Alerts**, **Syslog**, **Traps**, **Message Center**, **Reports**, **thwack**, **Custom Summary**, and **Training**.

Summary page

The **Summary** page is a simple view of the Orion NPM system, as well as an overview of your monitored network. By default, the **Summary** view in the Orion dashboard is the very first thing that will appear every time a user logs into Orion NPM. Some of the modules in the **Summary** page include a list of all monitored nodes, the network map, all triggered events, the last 25 events, and so on.

Groups page

There is also a **Groups** summary page. In Orion NPM, groups are a logical gathering of monitored nodes. For example, one group could include all nodes in one data center or building. Another group could include all nodes in a single geographical location. Groups are especially helpful for managed service providers as a group can be created for each customer's location. The options are virtually endless when it comes to groups.

Top 10

Top 10 is a page with all of the generic "Top 10" modules in one location. Some of the **Top 10** lists include **Top 10 Nodes by Percent Packet Loss** and **Top 10 Wireless Clients by Traffic** modules. This is a very helpful page that can help a network administrator or help desk technician prove that a device has been consistently problematic. It can even show if a device in a network has been "hogging" a great deal of bandwidth.

Events

The **Events** page is where you will find event log information about the Orion NPM system. An **event** is any type of change made to the Orion NPM system itself and may include a node threshold warning, Orion licensing warnings, and when anyone made a system change to Orion NPM. For example, if I was to add a new switch to Orion NPM to monitor, or change an interface's bandwidth threshold from a monitored node, or temporarily stop monitoring a node, these would display as events. Events are used primarily for internal auditing reasons and allow you to keep an eye on who, what, when, and where someone made a change to Orion NPM itself.

Alerts

The **Alerts** page is dedicated to the **Triggered Alerts for All Network Devices** module. In Orion NPM, an alert is some type of administrator notification about a node via MMS text message, e-mail, or other means. For example, if a monitored router stops responding to SNMP requests by Orion NPM, then an alert would be sent to an administrator about the situation.

Syslog

The **Syslog** page displays any type of syslog message captured by Orion NPM from all monitored nodes. If you do not have syslog configured on any of your devices, then there will be no data displayed in this page whatsoever.

Traps

The **Traps** page displays all current and historical SNMP Trap messages sent to Orion NPM by a monitored node. Essentially, **traps** are notifications sent to a network monitoring system from a monitored node about a significant event that impacts the device's normal working conditions. Provided a device is configured to send traps to a network monitoring system, the monitored node will do so when such an event occurs. Some examples would be, when a cooling fan fails, or if an uninterruptable power supply unit switches its breaker to battery mode. You can create Orion NPM alerts, create reports, and keep a history of trap messages received.

Message Center

The **Message Center** is the "master page" for every type of event, alert, syslog message, and trap message for the entire SolarWinds Orion NPM system. Essentially, the **Message Center** page is a combination of the **Events**, **Alerts**, **Syslog**, and **Traps** home pages.

Why is there a Message Center as well as Events, Alerts, Syslog, and Traps in the Home bar?

Because you may not want to allow certain users access to the complete Message Center, or you may only need to view a specific type of message without having to filter through the message types. You can configure access to specific users or user groups when they are available as different pages and modules.

Reports

Orion NPM can produce a great deal of reports about your networks, devices, and servers. The **Reports** page is where all of the reports are generated and downloaded from. A few reports that can be generated from this page are inventory, historical device status, device availability, and node downtime.

thwack

thwack is a forum-based site where thousands of SolarWinds customers and fans can talk about anything related to SolarWinds' products, especially Orion NPM. Many Thwack users share their own custom scripts that can be added to Orion NPM which enhance Orion NPM's functionality, and many of SolarWinds' software engineers and product managers participate in the Thwack community forums and respond to user issues and enhancement requests. It is here where they also offer their own tips and tricks about SolarWinds products as well. The link in the **Home** bar points directly to the Orion NPM section of the Thwack community site.

Custom Summary

By default, **Custom Summary** is a blank page and is intended to give Orion administrators a starting point for building custom monitoring pages.

Training

The **Training** page provides links directly to several of SolarWinds' free online training for Orion NPM. Some of the training links on this page are videos on SolarWinds' YouTube channel, others are links to articles in the Thwack community, and some are links to SolarWinds' online knowledgebase. The **Training** page is not a holistic training program for Orion NPM. Instead, it is a "getting started" point for people that are new to SolarWinds Orion NPM.

NETWORK tab

The **NETWORK** tab displays modules and tools specifically for network monitoring. Network engineers and server administrators use this tab the most.

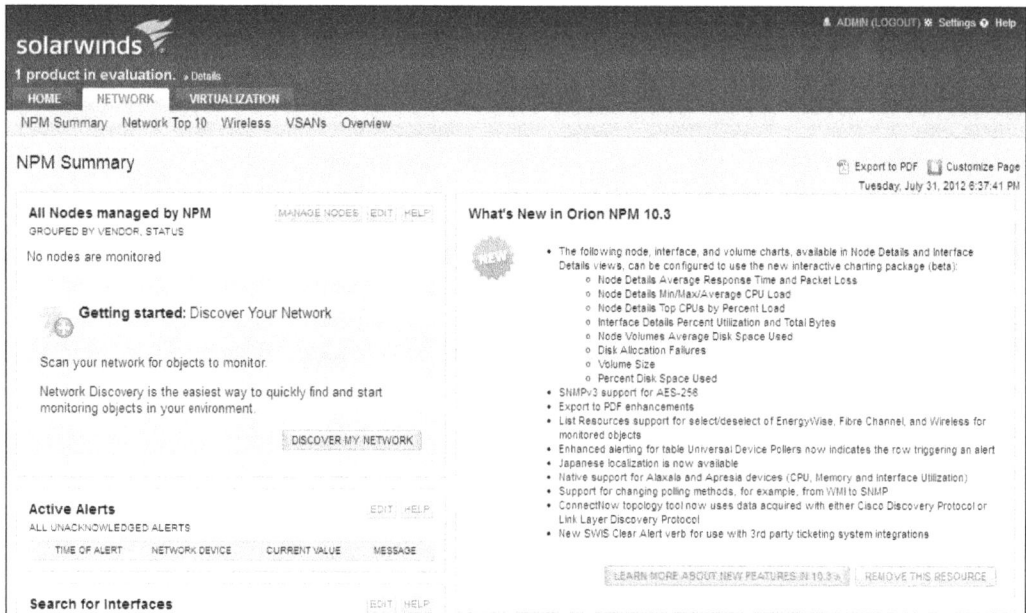

The link bar pages in the **NETWORK** tab are **NPM Summary**, **Network Top 10**, **Wireless**, **VSANs**, and **Overview**.

NPM Summary

The **NPM Summary** page is specific to network nodes that are being monitored by Orion NPM. The default modules on this page include all nodes managed by NPM, all active alerts, the Orion Network Atlas map, interfaces with high percent utilization, high errors and discards today, and a list of the last 25 events.

Network Top 10

The **Network Top 10** page is similar to the **Top 10** page of the **HOME** tab, but adds a few more new top 10 modules that are specific to displaying network-related information only. Some of the top 10 modules in this page include top 10 interfaces by traffic, top 10 volumes by disk space, and top 10 nodes by current response time.

Wireless

Orion NPM can monitor wireless access points from a variety of vendors and provide a detailed analysis for each access point and wireless controller. The **Wireless** page provides information on access point names or descriptions, IP addresses, types, list of SSIDs, channels, and how many clients are connected to which access points including the signal strength.

Due to the wide variety of vendors that make wireless access points and wireless controllers, only some of the most popular products from Cisco, Meru Networks, HP, Aruba, and Orinoco are supported out-of-the-box in Orion NPM. However, you can create Universal Device Pollers for access points and wireless controllers to monitor them in Orion NPM. Universal Device Pollers are covered in detail in a later chapter.

> For a current list of supported wireless devices and vendors see http://knowledgebase.solarwinds.com/kb/questions/4037/Wireless+OIDs+Polled+in+Orion+NPM.

VSAN

Storage Area Network (SAN) administrators can gain insight on how much traffic is passing through each device and interface from the **VSANs** page. Fiber channel, fiber channel over Ethernet, iSCSI, and many other SAN storage types are monitored specifically from this page.

Overview

The network **Overview** page is a very simple view of the entire list of monitored nodes and interfaces. Each interface is matched up to its corresponding node in the display grid. The **Overview** page displays only the "up" or "down" status of each node and interface that is monitored by Orion NPM.

Virtualization Summary

The link bar in the **VIRTUALIZATION** tab has only one page, **Virtualization Summary**. The **VIRTUALIZATION** tab is specific to VMware ESX and ESXi host monitoring only.

Orion NPM can monitor the CPU, memory, and disk storage of any VMware host as long as that host is running ESX 3.5 or above. Orion NPM will also display a list of virtual machines running on each VMware host. Server administrators or virtualization engineers that manage the VMware environment will need access to the resources in this tab.

> The **VIRTUALIZATION** tab is only used for monitoring the hardware and software resources on a VMware ESX/ESXi host. If you wish to monitor the virtual machines in a VMware host, you must add it as a monitored node in Orion NPM.

Native monitoring for Citrix XenServer and Microsoft Hyper-V are not directly supported at this time, only VMware is. However, you can monitor each Hyper-V and XenServer host using their built-in SNMP services. For monitoring each individual virtual machine, you can do so by adding each virtual machine to Orion NPM as you would monitor any other Windows or Linux server using WMI or SNMP.

Summary

In this chapter, we finished our initial configurations of our Orion NPM system. We saw an overview of the entire Orion dashboard, as well as an overview of the Orion Website Administration console. We also took a deep dive in user authentication with individual accounts and Microsoft Active Directory, we looked at how to configure thresholds, and we discussed Orion NPM licensing.

Now that our initial configuration is complete, let's move on and learn how to add devices to Orion NPM so we can start monitoring them!

3
Device Management

It is very rare that you will find two IT administrators very much alike. Most of us are like night and day; we are all very different in many ways. Every administrator has their own preferences on how they want to manage and monitor their network. Some prefer to maintain *God-like* control over their environment and micromanage every aspect of their network, especially when deciding who, what, when, and where to monitor their devices; while others prefer to utilize automation tools to help them easily stay on top of changes and additions to their network. Thankfully, SolarWinds has acknowledged this fact and provides several methods and processes for administrators to manage and monitor their network using Orion NPM.

This chapter is all about how to manage your network nodes in Orion NPM. We will talk about how to add nodes to Orion NPM using both automated and manual processes, how to change monitoring configuration settings, how to edit Orion NPM's polling methods, and how to configure interface monitoring settings. We are also going to discuss different node management techniques, management methodologies, group and dependency management, and credentials management.

Orion NPM provides administrators with two methods to add nodes, interfaces, and volumes to Orion NPM. One method is to use the Network Sonar Wizard. A second method is to add resources manually using the dashboard. Both methods eventually arrive at the same goal while providing administrators with several options on what they feel most comfortable with. By the end of the chapter you will have learned the following:

- Orion Discovery Central
- Device polling methods and polling configuration
- Node management techniques
- Group and dependency management
- Credentials management

Discovery Central

Discovery Central is where you will find Network Sonar Discovery, which is the configuration utility for the Orion Discovery Engine (a core component of Orion NPM). Discovery Central is usually the first location that Orion administrators will go to create a discovery profile once they are ready to start monitoring devices in Orion NPM.

A discovery profile is one that has been created using the Network Sonar Wizard to set up network scanning rules and schedules. You can create as many scanning profiles as needed that all have different rules and different schedules. For example, you could create a profile that scans for VMware hosts on a specific subnet automatically, and a second profile could be created for scanning only for switches and routers on a second subnet, but the scan must be run manually. The options are virtually endless.

Network discovery

A key feature of SolarWinds Orion NPM is the ability to automatically scan your networks for resources and present its findings in the dashboard. This makes it easier for administrators to be able to find devices on the network that need to be monitored, as well as making it faster to set up a new Orion NPM system.

A network discovery scan does not automatically add nodes to Orion NPM for monitoring. Network discovery operates much like a high school study hall teacher; it takes attendance and writes the names down, that's it. After Orion knows what can be monitored, an administrator must still select what resources they want monitored with Orion NPM.

So at this point you may be asking, "So why do I even want to bother with network scanning? I can just add all of the nodes manually myself!". Yes, that is true. You can add nodes manually and completely ignore the network discovery features of Orion NPM. But, consider you have several hundred nodes that you want to monitor with Orion NPM. If you just installed a fresh Orion NPM system, you would need to manually enter every one of your nodes one-by-one until all of them are monitored by Orion NPM.

Another consideration that I would keep in mind is for those that have a fairly dynamic network environment where change is commonplace. It is very easy to forget to log into a network monitoring system to remove old equipment and/or add new equipment. Personally, I have forgotten to add nodes to my SolarWinds Orion NPM system multiple times! Also, perhaps someone "fat-fingered" an IP address when trying to add a node to Orion NPM. The possibilities on why you would want to seriously consider using network discovery are endless. It is up to you if you want to actually use the network discovery tools in Orion NPM. You can always manually add nodes if you do not want to use, or cannot use, automated network discovery. Network Sonar Discovery is not a mandatory feature and is only available if you want to use it.

> If you are not interested in learning about network discovery, or only want to manually add nodes to Orion NPM, skip to the *Manually adding nodes* section in this chapter.

Next up, we are going to talk about the **Network Sonar Discovery** page. It is the configuration web page for the SolarWinds Discovery Engine, one of the core programs of Orion NPM, and it is here where current discovery profiles are edited and new ones are created. To access the **Network Sonar Discovery** page, click on the **Settings** link on the upper-right corner of the dashboard to enter Orion Web Administration, and then click on the **Network Sonar Discovery** link under **Getting Started with Orion**.

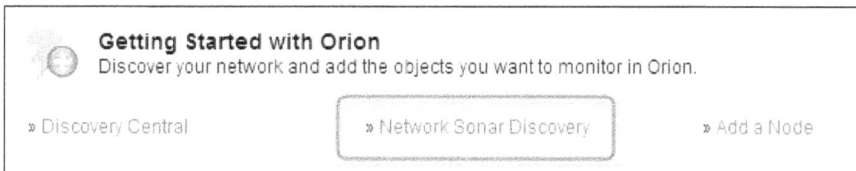

Getting Started with Orion
Discover your network and add the objects you want to monitor in Orion.

» Discovery Central   » Network Sonar Discovery   » Add a Node

Using the Network Sonar Wizard

The **Network Sonar Wizard** is a forms-based workflow wizard that makes it extremely easy to set up network scans. The wizard is used to create a network scanning profile. The workflow is always displayed on the top of the page so you can always see which step you are at in the wizard. You can always go back and make a change if you need to.

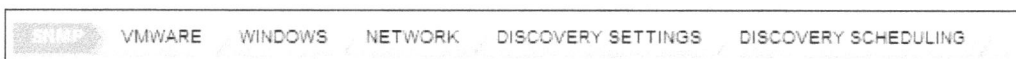

SNMP   VMWARE   WINDOWS   NETWORK   DISCOVERY SETTINGS   DISCOVERY SCHEDULING

Launching the **Network Sonar Wizard** is a fairly easy process. Simply click on the **Discover My Network** button to get started.

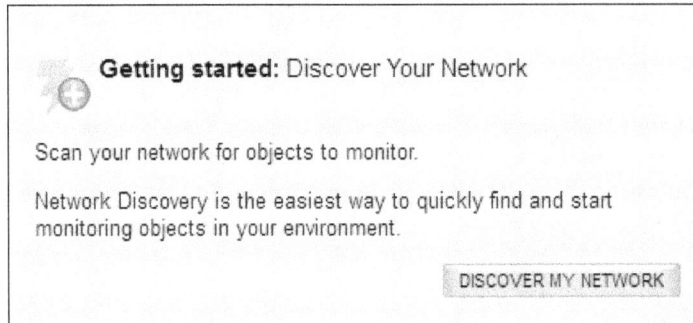

> **Getting started:** Discover Your Network
>
> Scan your network for objects to monitor.
>
> Network Discovery is the easiest way to quickly find and start monitoring objects in your environment.
>
> DISCOVER MY NETWORK

The following items are configured while creating a new discovery profile in the Network Sonar Wizard:

- SNMP credentials
- VMware user credentials
- Windows user credentials
- Network subnets, IP ranges, or singular network hosts
- Discovery customization settings such as timeout and retry values
- Network scan frequency and scheduling

SNMP credentials

The first step in the wizard takes you through adding your various SNMP Version 2 and Version 3 credentials. Normally, you would only need to add one or two SNMP credentials to a single discovery profile. However, you may need to add more depending on the scope of your discovery scan. It is possible that within a single subnet there may be different devices with different SNMP community definitions or credentials.

VMWARE   WINDOWS   NETWORK   DISCOVERY SETTINGS   DISCOVERY SCHEDULING

SNMP Credentials

Enter the SNMP credentials used on your network. The Discovery Engine automatically determines the community string and SNMP version to use for each network device. Credentials are used in the order listed below.

See more information about SNMP

➕ Add New Credential

Order	Credential	Version	Actions
1	public	SNMP v1 or v2c	⬆ ⬇ ✎ ✖
2	private	SNMP v1 or v2c	⬆ ⬇ ✎ ✖
3	solarwinds	SNMP v1 or v2c	⬆ ⬇ ✎ ✖

NEXT   CANCEL

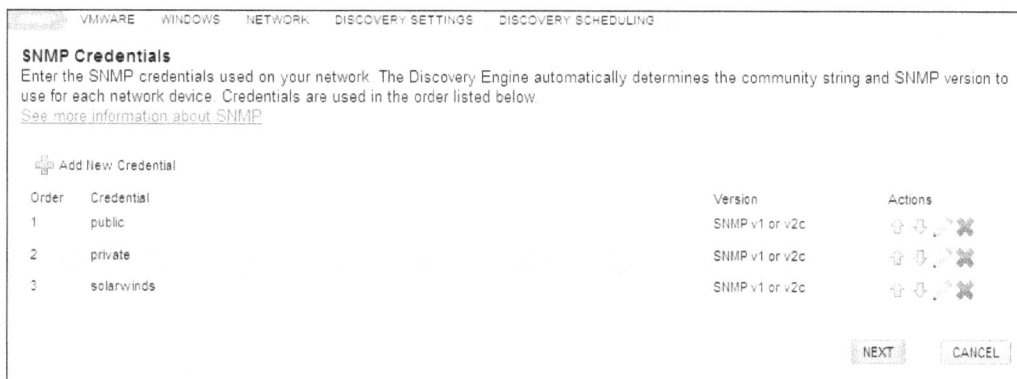

You can create as many discovery profiles as you want, with as many SNMP credentials as you like. Orion will save and store these credentials in its database so you don't have to re-type them every time you want to add a node or create a new scanning profile. However, if you do not need to enter any SNMP credentials, you do not have to do so. Simply click on the **Next** button to continue.

Once a node is detected during a network sonar discovery, Orion NPM will attempt to communicate with that node using all of the credentials defined in the profile. This means that if there are ten different SNMP credentials, Orion NPM will attempt all ten credentials in order. You can rearrange the credential priority as you see fit, and you can edit any of the credentials listed without leaving the wizard.

A best practice for SNMP credentials in a scanning profile is to include only what is necessary.

> Adding a large number of credentials to a discovery profile can make a network discovery job take a very long time to complete.

VMware user credentials

After the **SNMP** page comes the step to define VMware credentials. If you intend for your discovery profile to scan and discover VMware hosts, be sure to enter a VMware login ID that has permission to poll the host. Otherwise, you can skip this step.

SNMP | VMWARE | WINDOWS | NETWORK | DISCOVERY SETTINGS | DISCOVERY SCHEDULING

Local vCenter or ESX Credentials for VMware

Enter a local credential for the vCenter or ESX host server. Default ESX user name is "root".

Credentials are used in the order listed below.

☑ Poll for VMware

  Add vCenter or ESX Credential

Order	Credential		Actions
1	VMWare-Sonar		⇧ ⇩ ✖

BACK | NEXT | CANCEL

For VMware credentials, the exact same concepts regarding SNMP credentials apply here as well. Orion NPM will store these credentials in its database. It is highly recommended to *not* use the `root` user ID for security reasons. Create a different local vCenter or ESX credential with read-only access to the VMware host. If your VMware hosts are able to process LDAP or Active Directory credentials, then it is recommended to use one of these credentials.

If you do not need to poll for VMware hosts, then is it recommended to uncheck the box for **Poll for VMware**. Just as in SNMP credentials, you should only include the minimum VMware credentials in the scanning profile. Only include what is absolutely necessary.

Windows user credentials

Windows credentials are used to authenticate against Windows computers during a network scan.

SNMP | VMWARE | WINDOWS | NETWORK | DISCOVERY SETTINGS | DISCOVERY SCHEDULING

Windows Credentials
Enter the Windows credentials used on your network. Credentials are used in the order listed below. Learn more about Windows credentials

WMI is used to collect CPU, memory and volume data from Windows Servers that do not support SNMP, in addition to status, response time and packet loss.

  Add New Credential

Order	Credential		Actions
1	ORION-SONAR		⇧ ⇩ ✖

BACK | NEXT | CANCEL

The same concepts for VMware credentials and SNMP credentials apply here. (Are you starting to see a pattern emerge regarding credentials?). For Windows computers joined to a domain, it is recommended to create a domain service account, which will be used solely by SolarWinds Orion NPM to perform WMI queries. You may also use local Windows user accounts.

Network

The network step involves setting up the scanning profile to scan against one of three discovery methods:

- **Subnets**
- **IP Ranges**
- **Specific Nodes**

Subnets

One scanning method is by using subnets. When choosing to scan by subnet, you have two options. The first option is to define the *actual* subnet itself (that is `192.168.254.0/24`). When defining subnets, Orion NPM will scan for devices against every IP address within that subnet. Depending on the security requirements in your organization, normally it is safe to perform full subnet scans on class C networks. An example of a class C subnet is `192.168.1.0/24`.

In my experience, discovering devices by scanning an entire subnet is the most common option only when scanning `/24` subnets. This is a feasible option for those with smaller or more controlled networks, or if network nodes are spread across an entire subnet. To add a subnet, simply click on the **Add a New Subnet** button, type in your **Subnet** and **Subnet Mask**, and then click on **Save**.

The second option is to define a seed router. This is a very powerful option since instead of scanning an entire subnet, you can have Orion NPM poll for IP addresses stored in a router's ARP table. This is very useful if you have very large subnets (such as a class B or class A network) and you do not want to manually define every separate subnet per scanning profile.

You can add as many subnets and seed routers to the discovery profile as necessary. However, keep in mind that the more you add to the profile, the longer the network scan can take.

IP Ranges

Another scanning method is by using the IP range *within* a subnet, not the entire subnet itself. You would make this selection if you needed to scan against a specific range of IP addresses. Scanning by IP range helps to reduce the amount of time the discovery scan runs.

Many network administrators reserve some IP addresses within a subnet for network devices with static IP addresses, such as switches and routers. If you look at the previous screenshot, I am only scanning **192.168.1.200** to **192.168.1.254**. I am only scanning 54 IP addresses on this subnet because my DHCP server assigns dynamic DHCP addresses for workstations and printers in the 192.168.1.1 to 192.168.1.199 range.

Another example on why you would choose to scan a specific IP address range is if you had a class A network. Performing a network scan on the entire class A subnet could literally take hours for the scan to finish, so it would be wise to only select a portion of IP addresses in the discovery profile.

Specific Nodes

The last option available is **Specific Nodes**. Defining specific nodes allows an administrator to have complete control of what Orion NPM is allowed to scan. But this does mean that the administrator will have to manually change the discovery profile in the event of an IP address change on a device. If you have a strict list of IP addresses that you want to monitor, you know the specific IP addresses of specific nodes, or have strict security requirements to only scan what nodes are well known, this is the option you will choose.

SELECTION METHOD	One IP address or hostname per line
IP Ranges	192.168.1.250
Subnets	192.168.1.251
Specific Nodes	192.168.1.252
	192.168.1.253
	192.168.1.254

Validate

BACK   NEXT

Copy/pasting from a text file works best when adding specific IP addresses. You must enter one IP address per line in the textbox on this page.

> If you are intending on monitoring nodes with IPv6 addresses, you must choose the **Specific Nodes** method when creating a discovery profile. Alternatively, if you have a router with both an IPv4 and IPv6 address, you can use the seed router method to discover nodes with IPv6 addresses. Orion NPM cannot scan IPv6 addresses by IP range or subnet at this time.

Discovery Settings

The **Discovery Settings** step is where you will define the discovery profile name, description, and communication timeout values. In the following screenshot, notice the option at the bottom of the page for **Ignore nodes that only respond to ICMP (ping). Nodes must respond to SNMP, WMI**. If you do not want to include devices that respond to pings in the discovery, make sure you place a check mark in the box.

```
SNMP    VMWARE    WINDOWS    NETWORK    DISCOVERY SETTINGS    DISCOVERY SCHEDULING

Discovery Settings
Customize your network discovery by configuring the following settings.


    Name:          Nightly Scan
    Description:   Scans 192.168.1.0 at midnight


    SNMP Timeout:                                 3000  ms
    Search Timeout:                               2000  ms
    SNMP Retries:                                    1  retry(s)
    WMI Retries:                                     1  retry(s)
    WMI Retry Interval:                          10000  ms
    Hop Count:                                       0  hop(s)
    Discovery Timeout:                              60  min

                ☐ Ignore nodes that only respond to ICMP (ping). Nodes must respond to SNMP, WMI.
                » Learn more


                                              BACK    NEXT    CANCEL
```

All of the settings on this page are self-explanatory.

> Most desktop operating systems, network printers, and other consumer network devices will respond to ICMP requests, so it is a good idea to enable the **Ignore nodes that only respond to ICMP (ping). Nodes must respond to SNMP, WMI** option for heavily-populated networks.

Frequency and scheduling

The last step in configuring a discovery profile is for scheduling. There are only two different options available in this step, they are to set the frequency and execution.

In selecting the scanning frequency, you must define a custom time frame, either to run only once, or run daily. Running a scan daily is the natural choice for most small, medium, or more controlled networks. However, a custom scan may be needed to run weekly for less dynamic networks or every few hours for extremely dynamic networks, especially if there are constant network device changes and additions. The last frequency option is to run only once. This option is excellent for those who only want to execute Network Sonar Discovery manually instead of being on a regular scan schedule.

Finally, you must choose whether or not to execute the discovery now or later. Choosing to run now will save the profile and immediately execute the discovery scan. Choosing **No, don't run now** will only save and schedule the discovery profile. Choose the **Frequency** option **Once** and the execution setting **No, don't run now** to create the profile without scheduling it. The profile will then require an administrator to manually execute the scan.

Frequency:	Daily ⌄	Run this discovery every day at 04:00 AM
Execute immediately:	○ Yes, run this discovery now	
	● No, don't run now	

BACK   SCHEDULE

If you need to go back and make a change to your settings, or if you want to review your chosen settings, click on the **BACK** button. You can go back all the way to the first step of the Network Sonar Wizard if need be. Click on the **SCHEDULE** button in the final step to save the profile.

Next steps

Once a discovery profile has been created, access to the full Network Sonar Discovery page is enabled. You return here to view the results of the discovery scans and after Orion NPM has begun monitoring discovered nodes. To access the page, open Orion Web Administration and click on the **Network Sonar Discovery** link under **Getting Started with Orion**. Inside this page are the following three tabs:

- **Network Sonar Discovery**: This is where you will find all of your network scanning profiles
- **Scheduled Discovery Results**: This tab lists all of the resources that were discovered during its last scan
- **Discovery Ignore List**: This tab displays all of the previously discovered items that you have defined to be ignored in any future network scans

In **Network Sonar Discovery**, all of the network discovery scan profiles are listed onscreen. You can run a discovery job, schedule a discovery, change the settings of an existing discovery profile, create a new discovery profile, and delete a profile. The display grid shows the discovery profile title and description, its frequency, status, and the last time that the discovery was executed (if at all):

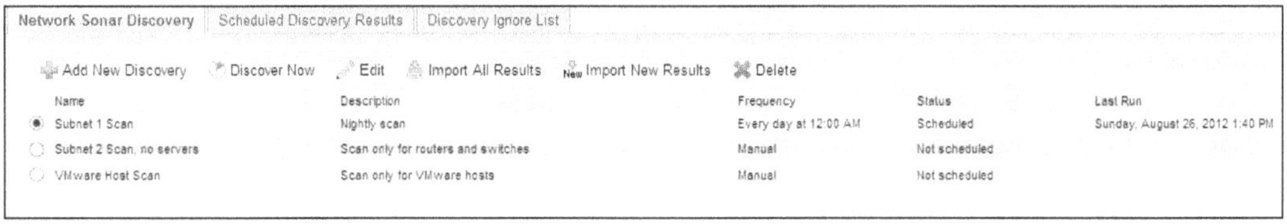

Network Sonar Discovery	Scheduled Discovery Results	Discovery Ignore List			
Add New Discovery	Discover Now	Edit	Import All Results	Import New Results	Delete
Name	Description		Frequency	Status	Last Run
Subnet 1 Scan	Nightly scan		Every day at 12:00 AM	Scheduled	Sunday, August 26, 2012 1:40 PM
Subnet 2 Scan, no servers	Scan only for routers and switches		Manual	Not scheduled	
VMware Host Scan	Scan only for VMware hosts		Manual	Not scheduled	

The buttons **Import All Results** and **Import New Results** are of special note here. Depending on what was discovered in the chosen scan, choosing **Import All Results** will simply add them to Orion NPM for monitoring, completely bypassing the requirement for an administrator to select what resources they want to add. The second button, **Import New Results**, performs the same task but only for new items discovered compared to the previous scan. Both of these options help to get a new Orion NPM system running quickly, but there are caveats with each option. An Orion administrator must always be aware of Orion NPM licensing when choosing what resources to monitor. This may not be a problem for those with an unlimited node count (SLX) license.

> You could easily use up all of your licensing if you import every single resource from a node!

If you want to execute a discovery scan immediately, choose the **Discover Now** button. A notification window similar to the following screenshot will be displayed and Orion NPM will perform the scan now:

Hop 0: Discovering: 192.168.0.254

Overall Progress

Current Phase

Nodes Discovered   7
Subnets Discovered   0

CANCEL

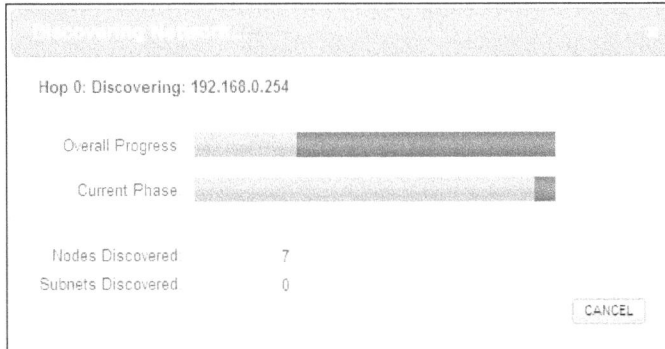

After an immediate discovery job finishes, the **Network Sonar Results Wizard** will display onscreen.

Adding nodes

Adding or importing nodes into Orion NPM is done in one of three ways: **Network Sonar Results Wizard**, **Add Node Wizard**, or **Scheduled Discovery**.

> You can access the same page at any time via Orion Web Administration by clicking on the **Network Sonar Discovery** link under the **Getting Started with Orion** module.

Adding nodes from a Scheduled Discovery

When a Scheduled Discovery has finished, the dashboard will display a notification in the top menu bar if Orion NPM found new nodes or node changes (such as a new interface or new hard disk).

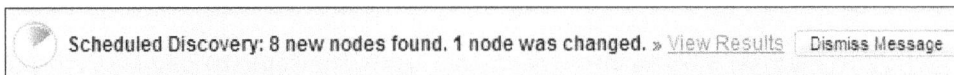

Scheduled Discovery: 8 new nodes found. 1 node was changed. » View Results   Dismiss Message

Clicking on the **View Results** link in the menu bar notification will open the **Scheduled Discovery Results** tab from the **Network Sonar Discovery** page.

Nodes that have been discovered are shown in the **Scheduled Discovery Results** page. The results page is fairly straightforward and it is relatively easy to see what Orion NPM discovered during a scan. The node's **Name**, **IP Address**, **Status**, **Description**, **Machine Type**, **Date Found**, and the profile used to find the node (**Discovered By**) are all displayed in the grid.

The **Status** column may be the most important piece of information in this tab since it shows if a node is newly discovered, if something has changed inside of a monitored node (such as an IP address change or interfaces have been added to it), or if the node was not added to Orion NPM after a previous discovery scan completed.

> A special note about changes: If a change is found on a currently monitored node, Orion NPM will continue to monitor that node as it was originally configured in the dashboard. An administrator must access the node's settings in Orion NPM in order to make a change to the monitoring configuration for that node. If this sounds confusing, here is an example: A switch was added to Orion NPM last month. Yesterday, a new trunk interface was added to the switch's configuration. Orion NPM will continue to monitor only what it was originally configured to monitor. If you want to monitor that new trunk interface, you need to edit the resources for the switch in the Orion dashboard.

Including nodes

Imported nodes that have been discovered during a network scan will be displayed in the **Scheduled Discovery Results** tab. First, choose the nodes that you wish to import into Orion NPM, and then follow the import workflow to configure and monitor the nodes. Perform the following steps in order to add a single discovered node to Orion NPM:

1. Click on **Import Nodes**.

2. The **Network Sonar Results Wizard** will be displayed. In the **DEVICES** step, place a check mark next to the nodes you wish to add to Orion NPM and click on **Next** to proceed to the **INTERFACES** step.

3. Choose the interfaces you wish to monitor and click on **Next** to proceed to the **VOLUMES** step.

Network Sonar Results Wizard

DEVICES   **INTERFACES**   VOLUMES   NPM IMPORT SETTINGS   IMPORT PREVIEW   RESULTS

Interface Types to Import
Select the interface types to monitor.

☑	Count		Interface Type
☑	2	↺	Loopback
☑	10	ᴍᴍ	Ethernet
☑	1	ᵢₗᵢ	Proprietary Virtual

BACK   NEXT

Pay attention to the list of interfaces in this step. Only select what makes sense. For example, if you do not need to monitor loopback interfaces, then make sure you remove the check mark for those interface types.

4. Choose the volumes to monitor, if any, and then click on **Next**.

Network Sonar Results Wizard

DEVICES   INTERFACES   **VOLUMES**   NPM IMPORT SETTINGS   IMPORT PREVIEW   RESULTS

Volume Types to Import
Select the volume types to monitor.

☑	Count		Volume Type
☐	1	🖴	Removable Disk
☑	1	🗄	Fixed Disk
☐	1	◎	Compact Disk
☑	1	🗇	Virtual Memory
☑	1	🗇	RAM

BACK   NEXT

Volumes primarily lie inside of servers or enterprise network appliances (such as load balancers, packet shapers, and NACs) and are usually not installed in routers or switches. However, times are changing and it is not uncommon to find some type of enterprise class router or switch chassis with its own disk volume(s). Remember that each volume will use one license in Orion NPM. Unless you absolutely must monitor a removable disk such as a DVD-ROM, never monitor a removable disk drive.

5. In the **NPM IMPORT SETTINGS** step, choose which interfaces you want to import and click on **Next**.

DEVICES	INTERFACES	VOLUMES		IMPORT PREVIEW	RESULTS

NPM Import Settings
Configure the following import settings.

Interfaces with the following status will be imported:

☑ Operationally Up

☐ Operationally Down

☐ Administratively Shutdown

| BACK | NEXT | CANCEL |

The default options that are selected in this step are all interfaces that are operationally up, while anything shut down or operationally down are skipped. This is appropriate for most situations, but you may have a need to include certain interfaces that might not be operational at the moment. Above all, be mindful of your licensing in this step.

6. The **IMPORT PREVIEW** step is the overview screen for all of the decisions made in the previous steps. If you chose to import multiple nodes, then you will see all of the nodes listed here. Otherwise, if you only imported a single node, only one will be listed. The **IGNORE** button at the bottom of the page is useful if you need to skip importing any nodes listed in this step. Ignoring nodes will simply remove them from this list. Click on the **IMPORT** button to start the import process.

Network Sonar Results Wizard

DEVICES	INTERFACES	VOLUMES	NPM IMPORT SETTINGS		RESULTS

Import Preview - JD-SOLARWINDS-2
Select devices, interfaces, and volumes that you wish to ignore or import. All ignored items will be removed from this list and will not be found during any future network discovery, manual or scheduled. If you wish to ignore items, do so before importing

		IP Address	Name	Machine Type	Volumes	Polling Method	Interfaces
☑							
☑		192.168.1.230	JD-SERVER1-2003	Windows 2003 Server	Fixed Disk, Virtual Memory, RAM	SNMP	Loopback, Ethernet
☑		192.168.1.241	Procurve-2600	ProCurve Switch 2600-8 PWR		SNMP	Ethernet, Proprietary Virtual, Loopback

| BACK | IGNORE | IMPORT | CANCEL |

7. The **RESULTS** step displays the import process. It is important to pay attention to this screen because any problems during the import process will be displayed in a message on this screen. Clicking on **FINISH** will close the **Network Sonar Results Wizard.**

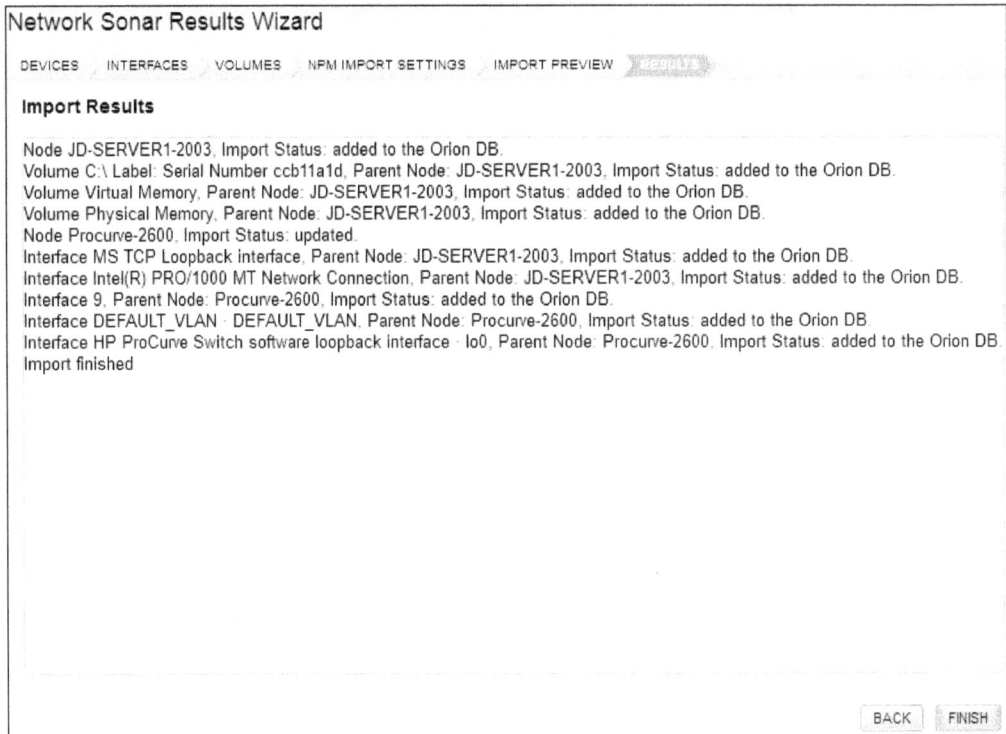

Network Sonar Results Wizard

DEVICES   INTERFACES   VOLUMES   NPM IMPORT SETTINGS   IMPORT PREVIEW   RESULTS

Import Results

Node JD-SERVER1-2003, Import Status: added to the Orion DB.
Volume C:\ Label: Serial Number ccb11a1d, Parent Node: JD-SERVER1-2003, Import Status: added to the Orion DB.
Volume Virtual Memory, Parent Node: JD-SERVER1-2003, Import Status: added to the Orion DB.
Volume Physical Memory, Parent Node: JD-SERVER1-2003, Import Status: added to the Orion DB.
Node Procurve-2600, Import Status: updated.
Interface MS TCP Loopback interface, Parent Node: JD-SERVER1-2003, Import Status: added to the Orion DB.
Interface Intel(R) PRO/1000 MT Network Connection, Parent Node: JD-SERVER1-2003, Import Status: added to the Orion DB.
Interface 9, Parent Node: Procurve-2600, Import Status: added to the Orion DB.
Interface DEFAULT_VLAN - DEFAULT_VLAN, Parent Node: Procurve-2600, Import Status: added to the Orion DB.
Interface HP ProCurve Switch software loopback interface - lo0, Parent Node: Procurve-2600, Import Status: added to the Orion DB.
Import finished

BACK   FINISH

Once the nodes have been imported, simply navigate to the **Orion Summary Home** screen where you can view its status. As you can see in the following screenshot, a router I imported now displays in the **Summary Home** page:

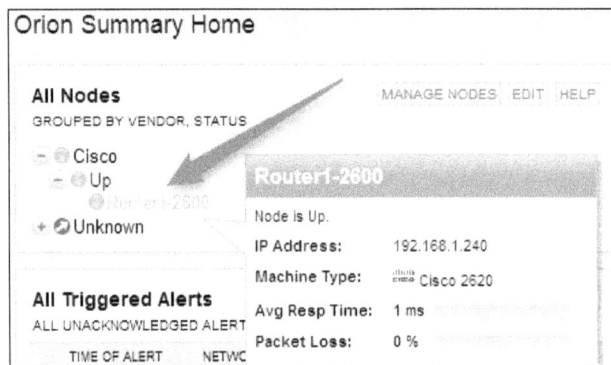

Orion Summary Home

All Nodes
GROUPED BY VENDOR, STATUS

MANAGE NODES   EDIT   HELP

- Cisco
  - Up
    - Router1-2600
  + Unknown

Router1-2600
Node is Up.
IP Address: 192.168.1.240
Machine Type: Cisco 2620
Avg Resp Time: 1 ms
Packet Loss: 0 %

All Triggered Alerts
ALL UNACKNOWLEDGED ALERT

TIME OF ALERT   NETWC

Discovery Ignore List

Orion network discovery scans can, and occasionally will, discover devices and nodes that you do not actually want to monitor with Orion NPM. If this happens, you will want to tell Orion NPM to ignore these devices in future scans. For example; Orion NPM has discovered a wireless desktop printer on my network. I do not want to monitor this printer whatsoever, so it needs to be added to the **Discovery Ignore List**.

The following is an example of how to load a node to the ignore list:

1. Place a check mark next to the node you wish to ignore.

   ☑ HPC510a   192.168.1.100 New Found

2. Click on the **Add to Ignore List** button.

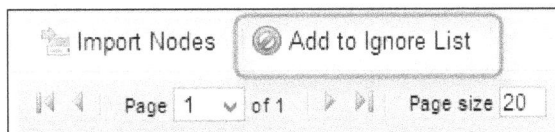

   Import Nodes   ⊘ Add to Ignore List

   Page 1 ∨ of 1   Page size 20

3. Click on the **OK** button on the notification window.

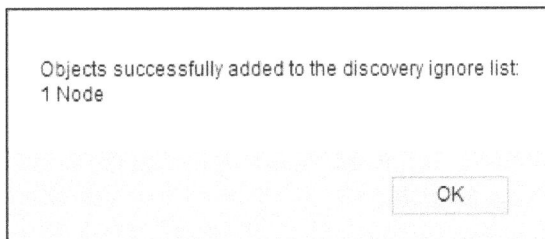

   Objects successfully added to the discovery ignore list:
   1 Node

   OK

4. Click on the **Discovery Ignore List** tab and verify that the correct node has been added to the list.

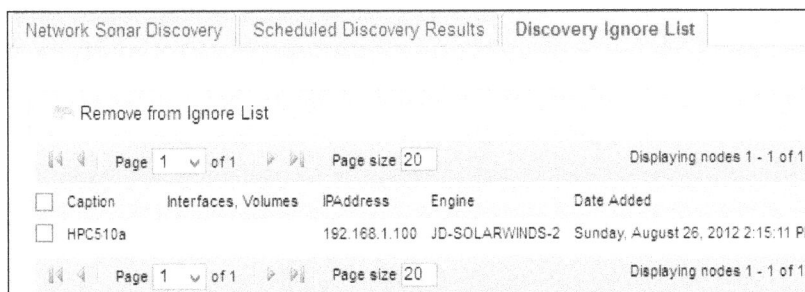

   Network Sonar Discovery | Scheduled Discovery Results | Discovery Ignore List

   Remove from Ignore List

   Page 1 ∨ of 1   Page size 20   Displaying nodes 1 - 1 of 1

   ☐ Caption   Interfaces, Volumes   IPAddress   Engine   Date Added
   ☐ HPC510a   192.168.1.100 JD-SOLARWINDS-2 Sunday, August 26, 2012 2:15:11 PM

   Page 1 ∨ of 1   Page size 20   Displaying nodes 1 - 1 of 1

Manually adding nodes

Performing network discovery scans are not the *only* way to import nodes into Orion NPM. Administrators can manually add nodes on-demand to Orion NPM. This is done using the **Add Node** wizard from the Orion Web Administration page.

Getting Started with Orion
Discover your network and add the objects you want to monitor in Orion.

» Discovery Central   » Network Sonar Discovery   » Add a Node

Add Node wizard

Once the **Add Node** wizard appears, simply fill in the blanks and follow the workflow to add the resource to Orion NPM. The workflow is always displayed at the top of the page in the **Add Node** wizard. A sample of the breadcrumb menu is shown in the next screenshot. You can always click on the **Back** button within any of the steps to go back and make changes:

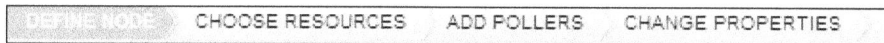

DEFINE NODE   CHOOSE RESOURCES   ADD POLLERS   CHANGE PROPERTIES

There are five steps involved when adding a node manually in Orion NPM. They are as follows:

1. Define the node's IP address or DNS hostname that you wish to monitor.

2. Choose a polling method.

3. Choose the elements you want to monitor from the node, such as interfaces, volumes, VLANs, PortChannels, CPU, RAM, or virtual memory.

4. Add a custom or universal poller.

5. Change the node's properties before finalizing the settings and then add the node to Orion NPM.

In the **DEFINE NODE** step, there are the following three things that you must do in order to proceed:

1. Define the IP address or DNS hostname of the node.
2. Choose one of four polling methods.
3. Configure the additional polling options for Cisco UCS or VMware.

When defining the **Hostname**, you can use the DNS hostname or the IP address of the node. Orion NPM assumes that the node has a fixed IP address assignment when an IP address is given. If you type an IP address, the **Dynamic IP Address** checkbox will grey out. If you have a node with a dynamic IP address (from **BOOTP** or **DHCP**), then you must type the DNS name of the node and check the **Dynamic IP Address** checkbox.

As shown in the following screenshot, you must also choose an appropriate polling method:

There are the following four different polling methods available and each is described, in detail, in the next section of this chapter:

1. **No Status**: Orion NPM will not collect any data on this node and it will not monitor it. It will only add the node to the summary list on the dashboard.

2. **Status Only**: This is for monitoring a node via ICMP (PING) only. Very little data can be gathered from this option.

3. **Windows Servers**: Orion NPM will poll the node using **Windows Management Instrumentation (WMI)** as well as ICMP.

4. **Most Devices**: This is the default option and is suitable for most devices. In it, you need to define the SNMP credentials, SNMP version, and port.

The final options in this step are to define **Additional Monitoring Options**. If you are monitoring **Cisco UCS** or **VMware** nodes, make sure to place a check mark in the appropriate box, then click on **Next** to continue.

We are now moving on to the step to add **Universal Device Pollers (UnDP)** pollers. UnDP is a unique feature to Orion NPM where you can select very specific SNMP OID's from your device to collect specific data from a monitored node. Universal Device Pollers are covered in detail in *Chapter 5, Network Monitoring II*. Once you have made your selection, click on **Next** to continue.

Add Node

DEFINE NODE   CHOOSE RESOURCES   ADD POLLERS   CHANGE PROPERTIES

Add UnDP Pollers to Router1-2600
Select universal device pollers to add to node

⊟ Example
  ☑ ciscoEnvMonFanState (The current state of the fan being instrumented.)
  ☑ ciscoEnvMonSupplyState (The current state of the power supply being instrumented.)
  ☐ ciscoEnvMonTemperatureStatusValue (The current measurement of the test point being instrumented.)
  ☐ ciscoEnvMonTemperatureStatusValueFahrenheit (The current measurement of the test point being instrumented, in ...)
  ☐ upsAdvBatteryCapacity (The remaining battery capacity expressed in percent of full capac...)
  ☐ upsAdvInputLineVoltage (The current utility line voltage in VAC.)
  ☐ upsAdvOutputLoad (The current UPS load expressed in percent of rated capacity.)
  ☐ upsBasicBatteryStatus (The status of the UPS batteries A batteryLow(3) value indicates ...)
  ☐ vmGuestOS (Operating system running on this vm.)
  ☐ vmGuestState (Guest operating system ON or OFF.)

BACK   NEXT   CANCEL

The final step is **CHANGE PROPERTIES**. It should look identical to the **DEFINE NODE** step aside from two additional properties at the bottom of the page: **Polling** and **Custom Properties**.

Polling

Node Status Polling: 120   seconds
Collect Statistics Every: 10   minutes
Poll for Topology Data Every: 30   minutes
Polling Engine: ◉ SOLARWINDS (Primary)

Custom Properties

City: Orlando, FL
Comments: Test lab 2600 router
Department: Buzzinga

BACK   OK, ADD NODE   CANCEL

The **Polling** area is where you define the polling engine settings for this specific node. The default setting is to check the heartbeat of the node every **120** seconds, collect various statistics from the node every **10** minutes, and poll for network topology data every **30** minutes. The **Custom Properties** area is where you can add some administrative notes about the node. After you have finalized the settings, click on **OK, ADD NODE** to import it into Orion NPM. Once the node has been added, you can find it on the **Orion Summary Home** page under the **All Nodes** module.

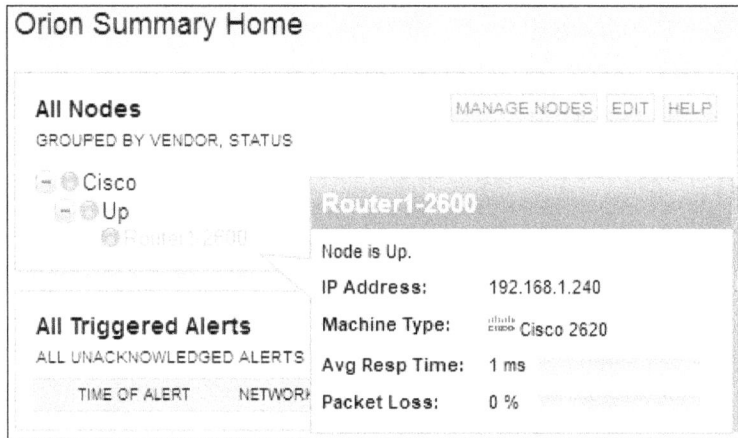

Orion Summary Home		
All Nodes		MANAGE NODES EDIT HELP
GROUPED BY VENDOR, STATUS		
⊟ ● Cisco	**Router1-2600**	
⊟ ● Up	Node is Up.	
● Router1-2600	IP Address:	192.168.1.240
	Machine Type:	cisco Cisco 2620
All Triggered Alerts	Avg Resp Time:	1 ms
ALL UNACKNOWLEDGED ALERTS		
TIME OF ALERT NETWORK	Packet Loss:	0 %

Polling

Most network devices and servers use counters and gauges to track performance and network activity against a variety of hardware and software resources inside of the node. The data from each of these counters and gauges can be extracted by Orion NPM using industry standard protocols such as WMI and SNMP and presented in an easy-to-understand format through the Orion dashboard. The process for extracting the counter and gauge data from a node is called **polling**. There are two reasons why you poll nodes, they are as follows:

- To verify the node's status
- To gather statistics

Orion NPM polls for status to know if the node is available or if there is a problem. A common node status is if it is up or down. If the node is *up*, it is usually operating normally. If the node is *down*, then there is a good chance there is a problem with that device.

The second reason why we poll nodes is to gather statistics from the node's counters and gauges. We can gather CPU usage, free space available on storage volumes, how many errors have been countered on a network interface, and much more.

Before a node is imported into Orion NPM, the software needs to know how it will be able to communicate with a node and what data it will be able to extract. Beyond the web-based dashboard, alerting engine, and other features of Orion, the polling engine is what truly drives a big portion of the Orion NPM system.

The following is involved with the polling process:

- Defining polling methods and polling settings
- Selecting resources/data to gather
- Collecting interface statistics
- Selecting an appropriate polling engine

Polling methods

The four polling methods in Orion NPM are as follows:

- **No Status: External Node**
- **Status Only: ICMP**
- **Windows Servers: WMI and ICMP**
- **Most Devices: SNMP and ICMP**

The first is **No Status: External Node**, which means Orion NPM will not poll this node whatsoever and will not attempt to detect its up/down status. This option is for those that wish to add some type of node reference to the dashboard. Since Orion NPM isn't monitoring the status of this type of node, it is not using up a license.

The **Status Only: ICMP** option tells Orion NPM to only use PING requests and replies to detect the node's up/down status. Checking the ICMP status is one of the simplest ways to monitor any type of node.

The **Windows Servers: WMI and ICMP** option is strictly for polling CPU, memory, volume, and ICMP data from Windows-based computers. You cannot poll interfaces using the WMI and ICMP option. All Windows operating systems, including all desktop operating systems and server operating systems, have SNMP services embedded in its software, but they are disabled by default.

> If you want to poll for interface data on a Windows operating system with Orion NPM, you must enable and configure SNMP services on the Windows computer and choose the **Most Devices: SNMP and ICMP** polling option in Orion NPM. For instructions on how to set up SNMP services in a Windows Server, see the Microsoft TechNet article at `http://support.microsoft.com/kb/324263/en-us`.

You cannot use the WMI and ICMP option to poll Linux computers, Mac OS X computers, or network devices. This option is strictly for Windows computers. When choosing the WMI and ICMP option, you must enter the credentials that have access to perform WMI scans against the node. SolarWinds Orion will store those credentials in its database and bind them to the node specified. The username must always be in domain/username format. In the following example, I added the controller Windows server to Orion NPM using the IP address **192.168.1.220**. For the credentials, I used a domain service account with the username **Sonar**:

The final option, **Most Devices: SNMP and ICMP**, is the default selection, and it is what you will select 99 percent of the time you manually add a node to Orion NPM, especially if that node is a router or switch. The only exceptions to this rule are Windows computers.

The default **SNMP Version** selected is **SNMPv2c**, the **SNMP Port** selected is **161**, and the checkbox to **Allow 64 bit counters** is checked. You can change any of these settings as you see fit. However, it is highly recommended to use **SNMPv2c** or **SNMPv3** and leave the checkbox enabled to allow 64-bit counters.

> SNMP counters wrap (meaning they go back to zero) when they hit their maximum limit. 64-bit counters allow for more counters in high capacity network interfaces such as 1 Gigabit and 10 Gigabit NICs. If you are using SNMP v2c or v3 to monitor your devices, it is highly recommended to leave the **Allow 64 bit counters** checkbox enabled. 64-bit counters are only supported in SNMP v2c and v3, while SNMP v1 supports only 32-bit counters. For more information on counters, Cisco has an excellent FAQ document on the subject at `http://www.cisco.com/en/US/tech/tk648/tk362/technologies_q_and_a_item09186a00800b69ac.shtml`.

Hostname or IP Address:	192.168.1.240	IPv4 and IPv6 formats are both valid
	☐ Dynamic IP Address (DHCP or B.O.O.T.P.)	
Polling Method	○ **No Status: External Node** No data collection for this node.	
	○ **Status Only: ICMP** Collect status, response time and packet loss data only, using ICMP (ping).	
	○ **Windows Servers: WMI and ICMP** Collect CPU, memory, volume and ICMP data for Windows Servers that do not support SNMP.	
	● **Most Devices: SNMP and ICMP - Recommended** Collect CPU, memory, volume and ICMP data for most SNMP-enabled devices.	

SNMP Version:	SNMPv2c ⌄	SNMPv2c is used, by default, when SNMPv3 is neither required nor supported.
SNMP Port:	161	
	☑ Allow 64 bit counters	
Community String:	solarwinds	Press down arrow to view all
Read/Write Community String:		
	○ Test Successful!	
	Test	

If the node you wish to monitor is a Cisco Unified Computing System node, or if the node is a VMware ESX/ESXi host, you must check the corresponding box to successfully detect and poll these nodes for data. Failure to enter this information will cause the detection to fail for VMware hosts.

Additional Monitoring Options:	☐ UCS manager credentials
	If the node hosts an UCS manager, check to enter credentials
	☑ Poll for VMware
	The VMware API is used to collect host information. SNMP is used to collect statistics for both hosts and guest nodes

When choosing the **UCS manager credentials** option, a new box appears at the bottom of the page where you must define the UCS credentials.

UCS credentials

Port:		(port on which the UCS manager listens)
	☐ Use HTTPS	
User name:		
Password:		
	Test	

When choosing the **Poll for VMware** option, define a user account that has access to the vCenter or ESX/ESXi host in order to continue to the next step. For security reasons, do not use the default root username. Create a new username with read-only access to the VMware host or vCenter server.

VMware Polling Settings

vCenter or ESX credentials:	vmware-sonar ⌄	
Credential name:	vmware-sonar	
User name:	sonar	Default is usually root for ESX credentials
Password:	•••••••••	
Confirm password:	•••••••••	
	Test	

Selecting resources that Orion NPM can poll from a router or switch will be very different from the resources in a server. Due to fact that there are literally thousands of device types that can be monitored with Orion NPM, it is near impossible to show you every device type in an example. Instead, I am going to give three examples, all from extremely common devices that you will monitor with Orion NPM. These are routers, switches, and servers.

When choosing which resources to monitor, it is a good idea to always monitor the **CPU & Memory** and **Topology** information since they don't require any additional licensing to do so in Orion NPM.

The following screenshot is an example of resources from a Cisco router:

Notice the serial interfaces on this router are not in use or are administratively down, so we are not going to monitor those interfaces. Also, items that may say **Loopback** or **Null** indicate that they are loopback (127.0.0.1) interfaces. I never recommend monitoring a loopback interface unless absolutely necessary, as it will use up an additional interface license. I selected the **FastEthernet0/0** interface since that is connected and in use.

The second example is for switches. The following screenshot is from an HP Procurve switch:

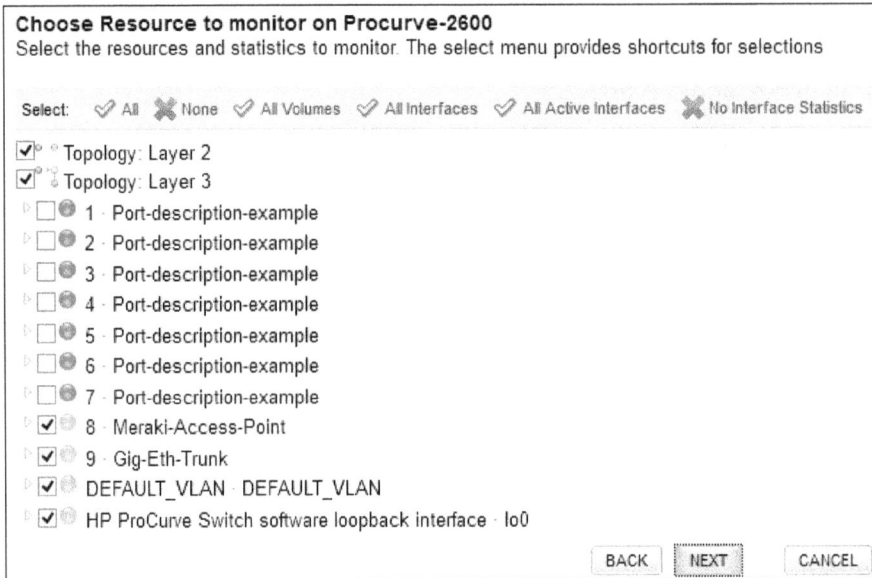

Choose Resource to monitor on Procurve-2600
Select the resources and statistics to monitor. The select menu provides shortcuts for selections

Select: ✓ All ✗ None ✓ All Volumes ✓ All Interfaces ✓ All Active Interfaces ✗ No Interface Statistics

☑ Topology: Layer 2
☑ Topology: Layer 3
☐ 1 · Port-description-example
☐ 2 · Port-description-example
☐ 3 · Port-description-example
☐ 4 · Port-description-example
☐ 5 · Port-description-example
☐ 6 · Port-description-example
☐ 7 · Port-description-example
☑ 8 · Meraki-Access-Point
☑ 9 · Gig-Eth-Trunk
☑ DEFAULT_VLAN · DEFAULT_VLAN
☑ HP ProCurve Switch software loopback interface · lo0

BACK | NEXT | CANCEL

The resources in a switch are similar to a router, but as you can see, there are many more Ethernet ports available that can be monitored. Orion NPM will automatically place a check mark next to every single active interface, so be aware of what is chosen. A good rule of thumb with selecting switch ports to monitor is to select only what you need. Doing this will help you keep your license usage as low as possible. The ports that you would want to consider monitoring are trunk ports, VLANs, PortChannels, and EtherChannels.

Finally, the following screenshot is an example of resources from a Windows Server when monitoring it via WMI and ICMP:

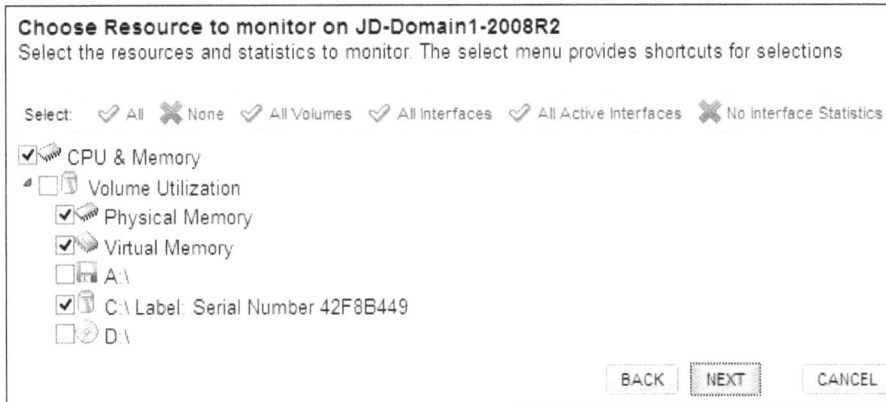

Choose Resource to monitor on JD-Domain1-2008R2
Select the resources and statistics to monitor. The select menu provides shortcuts for selections

Select: ✓ All   ✗ None   ✓ All Volumes   ✓ All Interfaces   ✓ All Active Interfaces   ✗ No Interface Statistics

☑ CPU & Memory
⊿ ☐ Volume Utilization
   ☑ Physical Memory
   ☑ Virtual Memory
   ☐ A:\
   ☑ C:\ Label: Serial Number 42F8B449
   ☐ D:\

[BACK] [NEXT] [CANCEL]

When compared to switches and routers, the resources in a server are almost completely different. The **CPU & Memory** resource still exists, but a new list of resources focusing on volumes is shown. In the example shown, three volume types have automatically been chosen; **Physical Memory**, **Virtual Memory**, and the Windows installation **C:**. If you remember the discussion about licensing in an *Chapter 2*, *Orion NPM Configuration*, this node will use three volume licenses.

> Network interfaces will not be available when you select the WMI and ICMP polling option at the time you add the server to the Orion NPM database. Orion NPM can only use SNMP to poll for interface data.

Interface statistics

When the SNMP and ICMP polling option is selected when adding a node to the Orion NPM database, there are several options available when monitoring interfaces. These options will not be available if the **ICMP Only** or **WMI and ICMP** polling options were chosen:

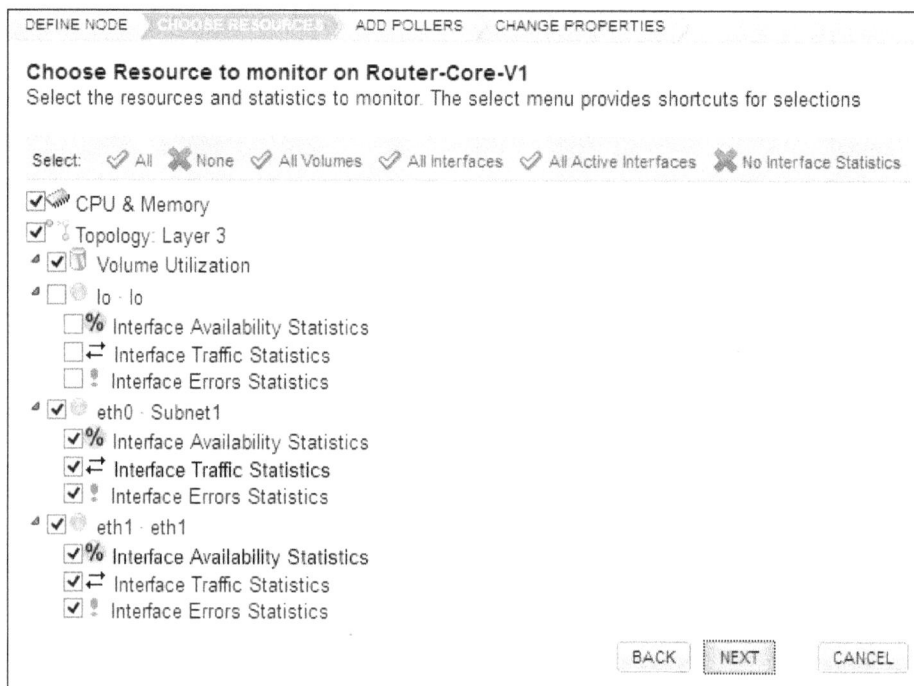

As shown in the previous screenshot, I have the ability to monitor three interfaces in this router. However, I am only interested in monitoring the **eth0** and **eth1** interfaces.

Expanding the interface display's settings gives us three options:

- **Interface Availability Statistics**
- **Interface Traffic Statistics**
- **Interface Errors Statistics**

The **Interface Availability Statistics** option polls the interface for its up/down availability and stores the historical data in the Orion NPM database. It is a no-brainer to leave this option selected since it is most likely that you want to know if an interface's availability is up or down and be able to create historical reports on that availability.

The **Interface Traffic Statistics** option polls the interface for response time, interface utilizations, total bytes transferred in an out, average amount of packets in and out, and other useful data. It is an excellent idea to keep an eye on the inbound and outbound traffic statistics to make sure that the interface isn't oversubscribed or undersubscribed. Therefore, it is recommended to poll for traffic statistics whenever possible.

The **Interface Errors Statistics** option polls the interface for packet loss, dropped packets (a.k.a. discards), and other interface error types. This statistic is especially useful when monitoring a WAN or wireless interface.

By default, all statistic options are automatically selected when you choose to monitor an interface. Doing so doesn't use any additional interface licenses and it is a good idea to capture all statistical information whenever possible. But with that said, there may be instances where you do not have a need to monitor traffic, errors, and/ or availability statuses. You can opt-out of gathering data from a node where you see fit. The Orion NPM polling process will run just a tiny bit faster, but not enough so that it is a large performance gain. If you are not sure what to do, simply keep all of them enabled.

Node & Group Management

Finally, we arrive at **Node & Group Management**. It is the dashboard location where every device that has been added to Orion NPM can be internally managed. It is here where devices are removed from Orion NPM, configurations are edited, polling settings are defined, interface properties are edited, and where groups and dependency groups are managed.

Nodes

Nodes and interfaces can be managed directly from the **Manage Nodes** page in Orion Web Administration. In it, you can edit the monitoring settings and configurations for each node.

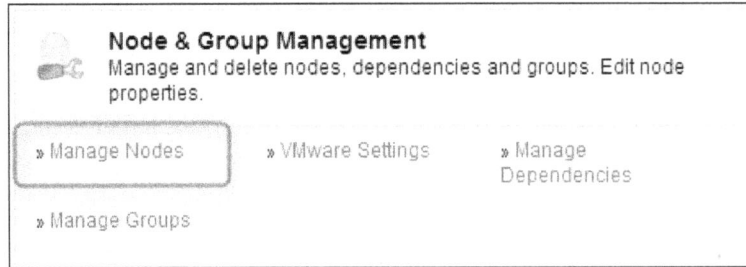

There are a variety of properties and options that can be edited from this page. You can edit a node's properties such as its name in the dashboard and the polling method, change the resources being polled by Orion NPM, un-manage a node, re-manage a node, and assign third-party universal device pollers.

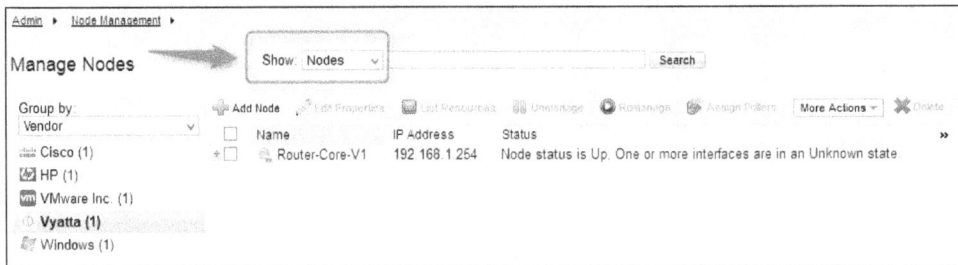

Editing the **Properties** of a node will open a page very similar to the **Add Node Wizard** which will allow you to change any of Orion NPM's properties it has stored for that node. If you needed to change the polling method, the device's name, or even the IP address of the node you are monitoring, you will need to edit the node's properties.

Changing the **Resources** will also take you through a page similar to the **Add Node** wizard, but will go straight to the **List Resources** step. Un-managing a node will disable Orion NPM from polling that device. When you re-manage the node, Orion NPM will resume polling as it has been configured to do so.

Universal device pollers are covered extensively later in this book, but the **Manage Nodes** page in Orion Web Administration is where you assign additional pollers to a node.

Groups

Orion NPM allows you to create custom groups. A **group** is a logical label of nodes or elements under one umbrella. Why would you want to use groups? Here are several examples:

- You are a **Managed Service Provider** (**MSP**) and manage and monitor all network devices and servers for your customers. For each customer, you create a group that includes all of their network equipment.

- Your business is headquartered in one area of the country, with several branch offices scattered throughout the world (each with network equipment monitored by Orion NPM). You can create a group in Orion NPM and add nodes for each location to their respective groups.

- A company has several buildings connected, each with several network closets, to your network over a large metropolitan area. You want to create a group for each building in Orion NPM.

- You have several WAN links to different Internet service providers on different routers in your network. You create a group to make it easier to display all of the interfaces on the Orion dashboard.

- You want to monitor several volumes in different servers on your network. You create a group that includes all of these volumes.

Another reason for building groups in Orion NPM is that, thanks to Orion NPM's robust user permissions options, you can limit user account access to nodes in specific groups. This is great if you are an MSP since you would be able to grant access for a customer to log into *your* Orion NPM system to view monitoring on *their own* devices! Or, in a business with distributed IT personnel, you can limit access to a group for a specific team of individuals. The options are virtually endless.

Creating and editing group memberships is done from the **Manage Groups** page under the **Node & Group Management** module in Orion Web Administration.

There are several options you need to choose when creating a group in Orion NPM. First, you must define a unique group name and type a description. You must also choose the **Status Rollup Mode** and **Status Refresh Frequency**.

The three options for **Status Rollup Mode** are **Show Best Status**, **Mixed Status Shows Warning**, and **Show Worst Status**. The default option is **Mixed Status Shows Warning**, and this is the recommended option.

To create a group, perform the following steps:

1. Open Orion Web Administration.

2. Click on the **Manage Groups** link under **Node & Group Management**.

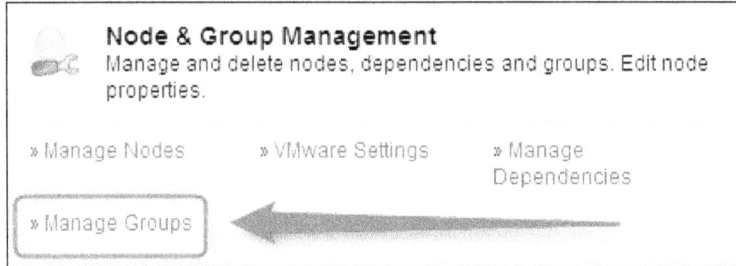

3. Click on **Add New Group**.

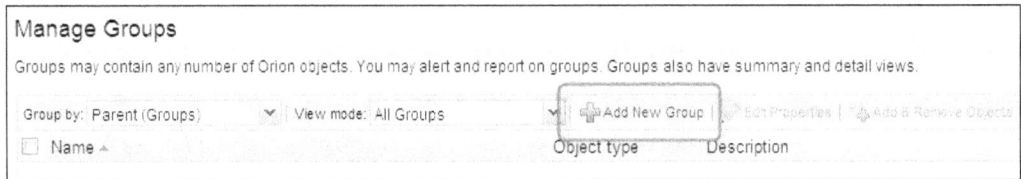

4. Type the group's **Name** and **Description**, and then click on **Next** to continue.

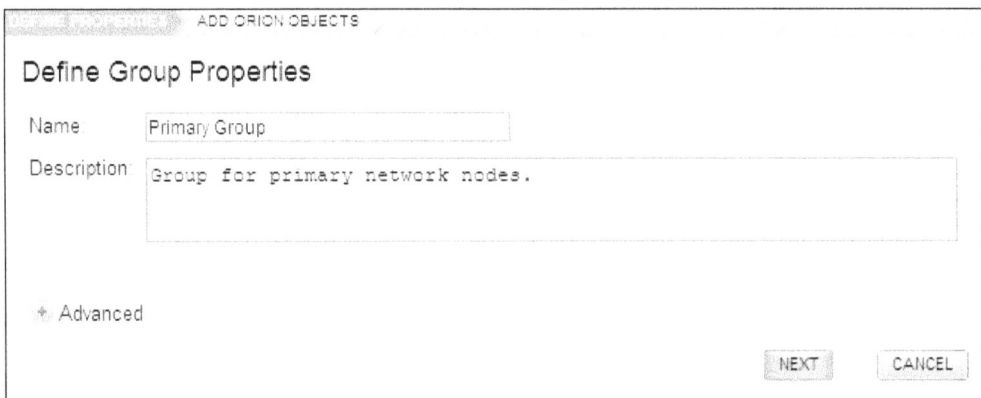

5. Every monitored node and interface that has been added to Orion NPM will be displayed in the **Available Objects** pane on the left-hand side. Choose all of the items that you want to add to the group, and then click on **CREATE GROUP** to continue.

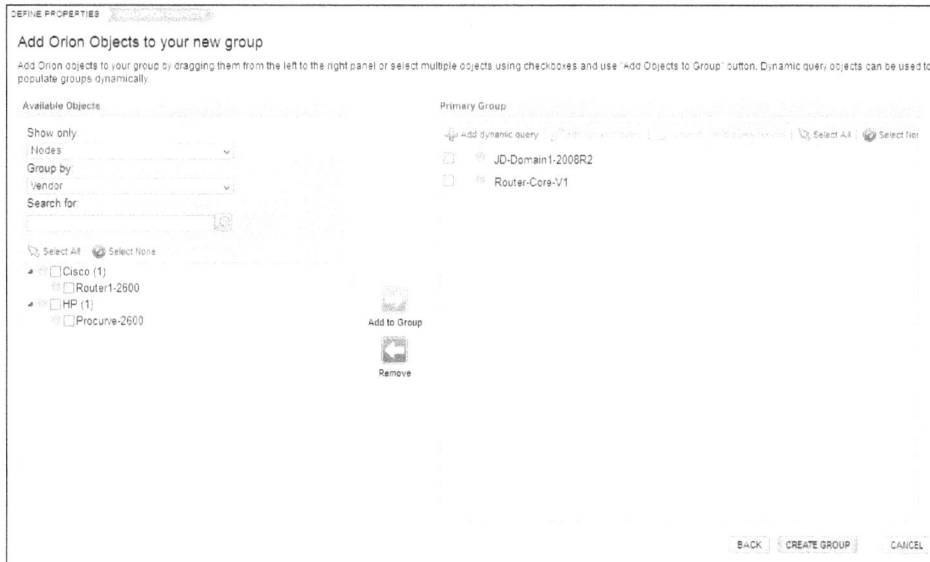

6. The new group will be listed on the **Manage Groups** page. Click on the caret to expand the group to see which nodes were added.

7. Groups will also be listed on the **Orion Summary Home** page under the **All Groups** module.

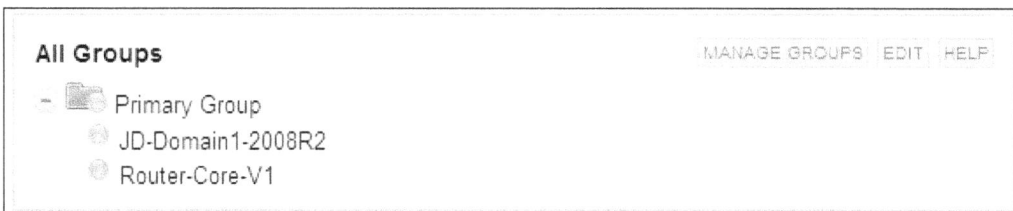

You can always edit a group's membership and settings from the **Manage Groups** page. One special thing to know about groups is that it is possible to nest groups inside of other groups.

Dependencies

A **dependency** is where you can define what node or interface is dependent on another. Dependencies operate in a parent-child relationship; when the top-level node goes down, Orion NPM automatically assumes that the child nodes are down as well and will mark them as **Unreachable** in the dashboard.

Dependencies are used mostly for alerting reasons. Orion NPM will only notify an administrator about the parent node or interface not responding instead of multiple alerts for multiple nodes. This helps with allowing administrators to get down to the bottom of a problem more quickly and will not inundate them with multiple alerts.

When you create a dependency, first you must choose what the parent will be. Then, you choose the child. The child can be a single monitored node or interface, or you can choose a node group or interface group.

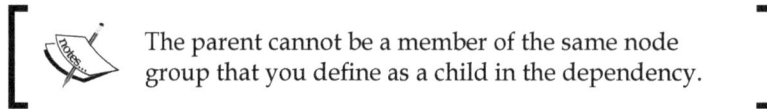

> The parent cannot be a member of the same node group that you define as a child in the dependency.

Creating dependencies is done from Orion Web Administration under the **Node & Group Management** module.

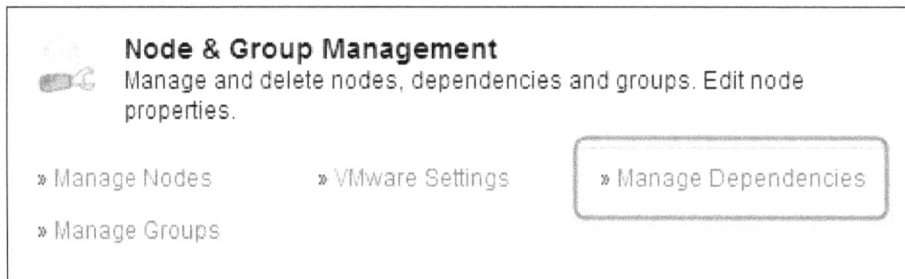

Node & Group Management
Manage and delete nodes, dependencies and groups. Edit node properties.

» Manage Nodes   » VMware Settings   » Manage Dependencies

» Manage Groups

Perform the following steps to create a dependency in Orion NPM:

1. Click on the **Add new dependency** button.

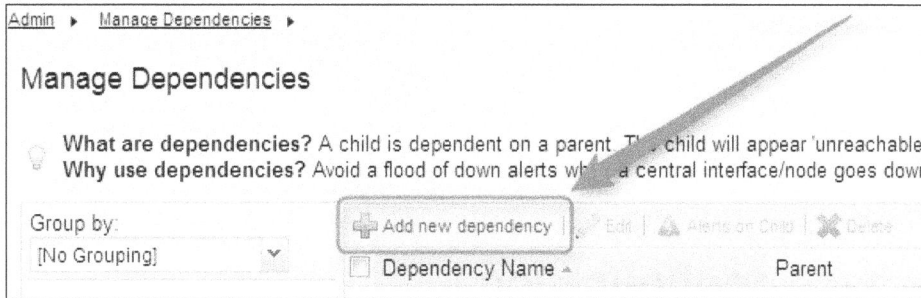

2. Choose the parent node, then click on **Next**.

3. Choose the child node, then click on **Next**.

4. Review the settings in the dependency, make sure to define a meaningful **Dependency name**, and then click on **Submit** to finish.

5. Your new dependency will now appear in the list in the **MANAGE DEPENDENCIES** page.

6. Dependencies are also displayed on the **Orion Summary Home** page under the **All Dependencies** module.

All Dependencies		MANAGE DEPENDENCIES EDIT HELP
Dependency Name	Parent	Child
Domain Controller	Router-Core-V1	JD-Domain1-2008R2
Router-Core-V1 dependency	Router-Core-V1	Router1-2600

Interface configuration

When you are monitoring a node using SNMP, and you are monitoring interfaces from that node, Orion NPM is able to detect the maximum bandwidth type of the interfaces that are a part of the node. For example, if you add a switch, and its uplink port is Gigabit Ethernet, then Orion NPM knows that the maximum throughput of that interface is 1000 Mbps. However, consider if you are using one of your Gigabit Ethernet ports as a WAN uplink to the Internet. What if the Internet service provider has a maximum download speed of 200 Mbps and an upload of 50 Mbps on this interface? Unless you configure the bandwidth options for that interface, Orion NPM will only report on the default settings.

To change the configuration settings for a monitored interface, you must first open the **Node Details View** page of the monitored node from the **Orion Summary Home** page, or from **Node Management** in Orion Web Administration.

> The settings defined in the **Edit Interface** page only applies to the polling configuration for Orion NPM and does not affect the operation of the interface itself.

From the **Node Details View** page, perform the following steps:

1. Scroll down to the **Current Percent Utilization of Each Interface** module and click on the interface you want to edit. In this example, I will be editing the **Gi0/2** interface settings.

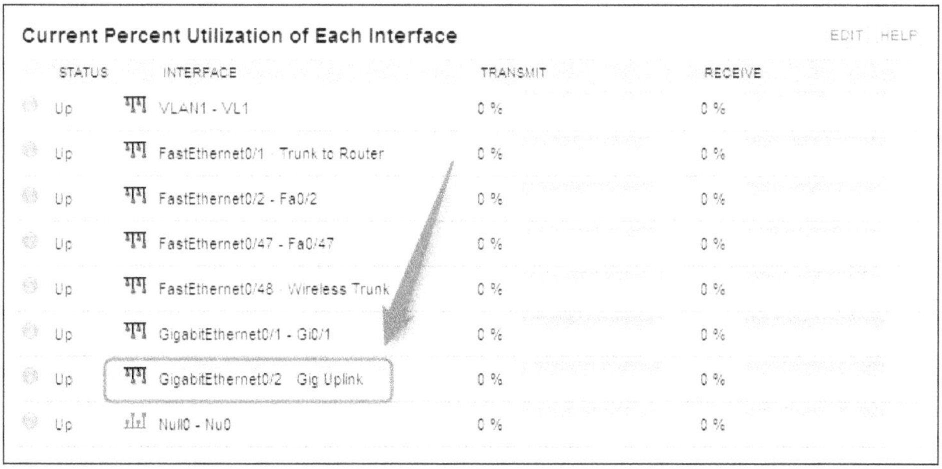

Current Percent Utilization of Each Interface   EDIT HELP

	STATUS	INTERFACE	TRANSMIT	RECEIVE
	Up	VLAN1 - VL1	0 %	0 %
	Up	FastEthernet0/1 Trunk to Router	0 %	0 %
	Up	FastEthernet0/2 - Fa0/2	0 %	0 %
	Up	FastEthernet0/47 - Fa0/47	0 %	0 %
	Up	FastEthernet0/48 Wireless Trunk	0 %	
	Up	GigabitEthernet0/1 - Gi0/1	0 %	0 %
	Up	GigabitEthernet0/2 Gig Uplink	0 %	0 %
	Up	Null0 - Nu0	0 %	0 %

2. Now, click on **Edit Interface** under **Interface Details**.

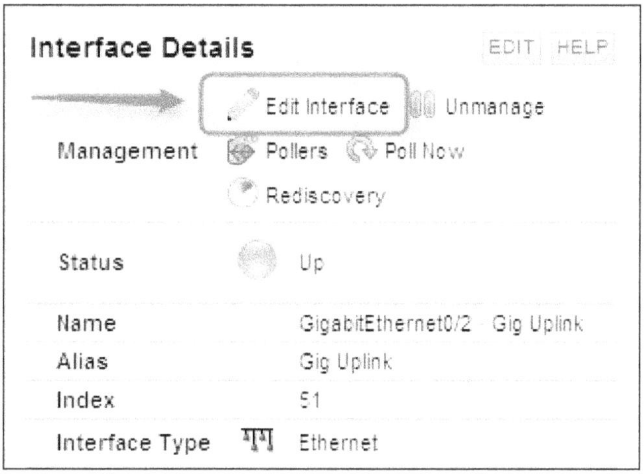

Interface Details   EDIT HELP

→ Edit Interface Unmanage

Management   Pollers Poll Now

Rediscovery

Status   Up

Name   GigabitEthernet0/2 Gig Uplink
Alias   Gig Uplink
Index   51
Interface Type   Ethernet

The **Edit Interface** page will appear. Available within it are five different settings:

- **General**
- **Bandwidth**
- **Polling**

- **Custom Properties**
- **Dependencies**

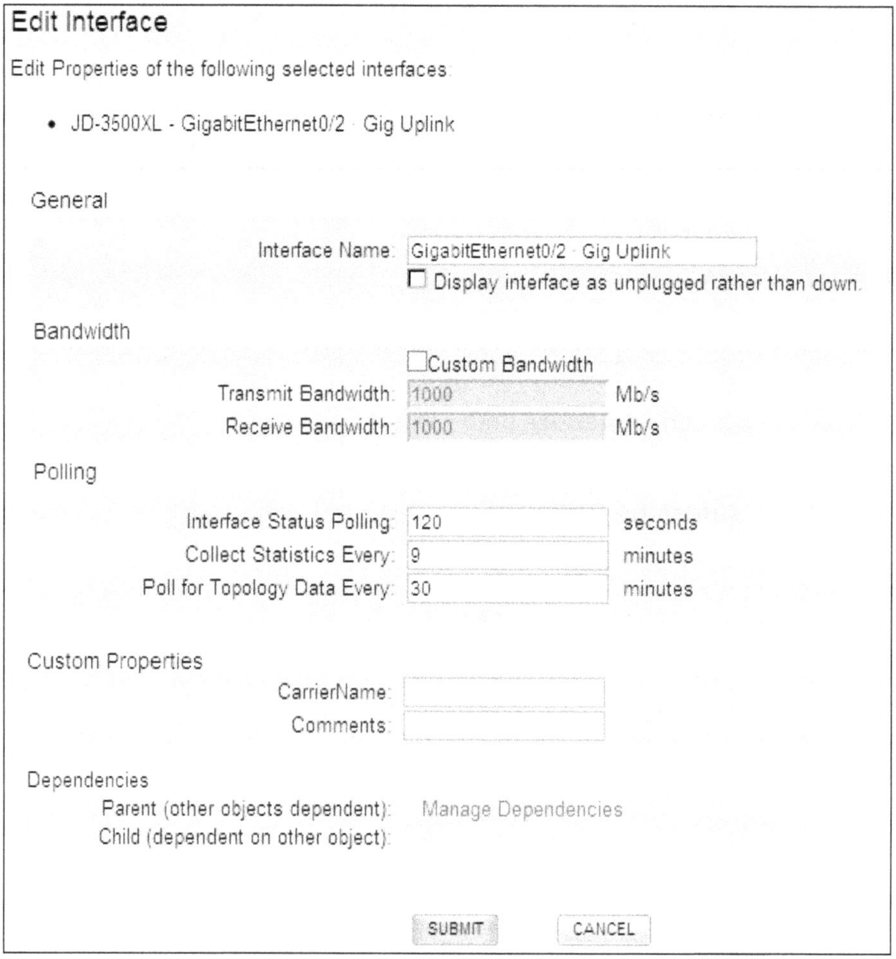

Edit Interface

Edit Properties of the following selected interfaces.

- JD-3500XL - GigabitEthernet0/2 - Gig Uplink

General

Interface Name: GigabitEthernet0/2 - Gig Uplink
☐ Display interface as unplugged rather than down.

Bandwidth

☐ Custom Bandwidth
Transmit Bandwidth: 1000   Mb/s
Receive Bandwidth: 1000   Mb/s

Polling

Interface Status Polling: 120   seconds
Collect Statistics Every: 9   minutes
Poll for Topology Data Every: 30   minutes

Custom Properties

CarrierName:
Comments:

Dependencies

Parent (other objects dependent):   Manage Dependencies
Child (dependent on other object):

SUBMIT   CANCEL

General

General options are where you can rename the interface and choose whether or not you want to display the interface as unplugged rather than down. The interface name is only an administrative note. If the interface already has a label assigned to it from the device (such as a port description in a Cisco switch), it will be displayed here by default.

The second option in the **General** section is to choose whether or not you want to display the interface as unplugged rather than down. By default, this option is left unchecked, which means that the Orion NPM alert engine will trigger an alert when the interface changes to a down state. Placing a check mark in this box will inform Orion NPM not to trigger an alert and instead mark the interface as unplugged. This option is useful when performing maintenance tasks on your network. Alerts are covered in more detail in *Chapter 6, Setting Up and Creating Alerts*.

Bandwidth

Bandwidth is the important part of this section. By default, both **Transmit Bandwidth** and **Receive Bandwidth** boxes are set to their maximum port speed and are grayed out. Check the **Custom Bandwidth** box to edit these areas. As stated before, you want to edit the bandwidth settings when monitoring WAN ports.

Polling

Polling options are inherited from the polling statistics for the node. However, Orion NPM allows you to edit the polling settings per interface.

Custom Properties

Custom Properties are tables added to the Orion NPM database for administrative and reporting purposes. You would use a custom property to make note of a carrier's circuit number for an interface, customer information, asset tag, location, or any other important information. **CarrierName** and **Comments** are the default custom properties available within the **Edit Interface** page. **Custom Properties** and the **Custom Property Editor** are covered in more detail in *Chapter 5, Network Monitoring II*.

Dependencies

We already discussed **Dependencies** in the previous section. However, Orion NPM provides a quick link to configure dependencies only for this interface. If there are already interface dependencies configured, they will be displayed here.

Credential libraries

When you import a node into Orion NPM and enter a set of credentials in order to poll that node, Orion NPM will store the credentials in its database. You can always go back and edit these credentials, add new ones, or delete them from the database. Orion NPM only stores the following three types of credentials:

- SNMP v3 credentials
- Microsoft Windows credentials
- VMware credentials

SNMP credentials library

If SNMP is chosen while adding a node to the Orion NPM database during the polling method, and you select SNMPv2 or SNMPv1, the community names are automatically saved to the Orion NPM database.

SNMPv3 is a more secure version and requires a user account, password, and other attributes. To accommodate this, Orion NPM will store all SNMPv3 credentials in its database and they will only be displayed under the **Polling Method** section within **Edit Nodes**. This makes it easier to reuse, add, or remove these credentials in the future, especially when adding more nodes. To view the SNMP credential library, open the properties of a node that is currently being monitored via SNMP and scroll to the SNMP settings for that node. As shown in the following screenshot, when the SNMP version is set to SNMPv3, the **Credential Set Library** will be displayed:

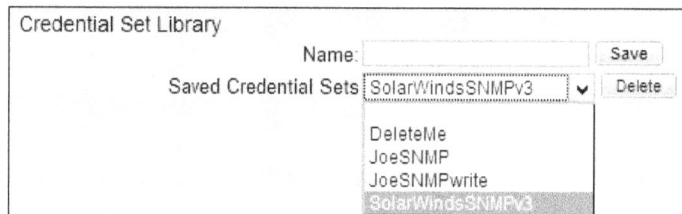

As you can see in the previous example, I have four different SNMPv3 credentials saved in my Orion NPM instance. Adding more SNMP credentials to the library is simple. Just fill out the form starting with the **SNMPv3 Username** and **Context**, **Authentication** type (**MD5** or **SHA1 hash**) or **Encryption** information, **Password**, and then type a **Name** for the credential. Finally, create an administrative name for the SNMPv3 credential to store it in the database, and then click on the **Save** button.

Deleting saved SNMPv3 credentials is simple as well. Choose a saved credential set from the drop-down list from any monitored node, and then click on **Delete**. Unfortunately, Orion NPM does not display how many nodes may be using a specific SNMP credential, so you will need to make sure that the SNMPv3 credential truly is not in use by a node manually before deleting it.

> When you delete an SNMPv3 credential, Orion NPM will not warn you if the credential is currently in use by a node and it will simply delete it. Do not delete an SNMPv3 credential unless you are absolutely sure it is not in use.

Windows credentials library

Windows credentials are found in the **Manage Windows Credentials** link in Orion Web Administration.

Every stored Windows credential will be displayed in the grid. All Windows credentials are shown in `domain\username` format as well as how many nodes have been assigned to a credential. From this page, you can create a new credential, edit a current credential, and delete any unused credentials.

If the Windows password has changed, you will need to update the credential in the **Manage Windows Credentials** page. Simply place a check mark next to the credential and click on **Edit Credential**. On the edit screen, update the information and click on **Ok**.

VMware credentials library

The last credential type is VMware user accounts, which you can find in the **VMware settings** link in Orion Web Administration. Click on the **VMware Credentials Library** tab to view the stored credentials.

VMware credentials operate in exactly the same fashion as Windows credentials. You can edit, delete, and add new credentials to the Orion NPM database. Unfortunately, Orion NPM also does not display how many nodes have a credential assigned to it.

Summary

So, that is it for this chapter! We covered quite a few topics. We learned how to use the network discovery automated tools to add nodes to Orion NPM, and how to manually add nodes. We learned all about polling, as well as how to manage the nodes themselves, their interfaces, and credential sets.

I'm going to keep the ball rolling and dive right into all of the details of monitoring in the next chapter. There is quite a bit of ground to cover, but I know that you will find it fairly easy to understand once we get started!

4
Network Monitoring Essentials

A great deal of content was covered in the previous chapters. We walked through installing an Orion NPM system, discussed network discovery, and added devices to our Orion NPM system. We are finally going to discuss using Orion NPM to actually monitor the performance of your network.

This chapter is actually one of the two parts. In part one, we are going to cover monitoring basics, customizing views, network device monitoring (such as routers and switches). In part two, we are going to focus directly on servers, virtualization, virtual storage area network (VSAN) monitoring, and customizing polling engines. By the end of this chapter you will have learned the following:

- Monitoring basics
- Customizing views
- Managing and adding views
- Editing monitoring resources
- Router monitoring
- Switches
- Wireless monitoring
- Wireless Controllers and Access Points

Monitoring basics

By now, you have already added some devices to your own Orion NPM installation and are ready to dive right in. In all actuality, network monitoring with Orion NPM fundamentally consists of two different actions. The first is Orion NPM polling devices and discovering nodes. The second is an administrator physically logging in to the Orion dashboard and looking at the statistics and node information.

All monitoring is performed by viewing the pages available from the various tabs at the top of the dashboard website. By default, the **Orion Summary Home** page opens directly after logging into Orion NPM's dashboard. This is the page you will find yourself looking at most of the time while managing and monitoring your network with Orion NPM.

There are several modules on every page of the dashboard that provide you with several pieces of information. If you are not sure what the content of a module contains or if you want more information about what the module is displaying, click on the **HELP** button at the top-right corner for a detailed explanation.

Map

The most prominent module displayed on the main page is the network map on the right-hand side. It is designed to display the "big picture" of your entire network that is monitored by Orion NPM. Questions that are quickly answered by the network map is, "Are some nodes up?", "Are some nodes down?", "Are there any network performance issues (slowness or packet loss) between WAN links?", or "Are there any network performance issues that I should be aware of?" The network map can point you in the right direction quickly if a network issue needs to be resolved. A picture truly does speak a thousand words!

The map that is displayed after a fresh installation of Orion NPM is the sample map. The sample map is only a placeholder and does not display any of your nodes on it. To create and customize your own network maps, you need to use the **Orion Network Atlas** utility. Creating network maps is covered in detail in *Chapter 7, Producing Reports and Network Mapping*.

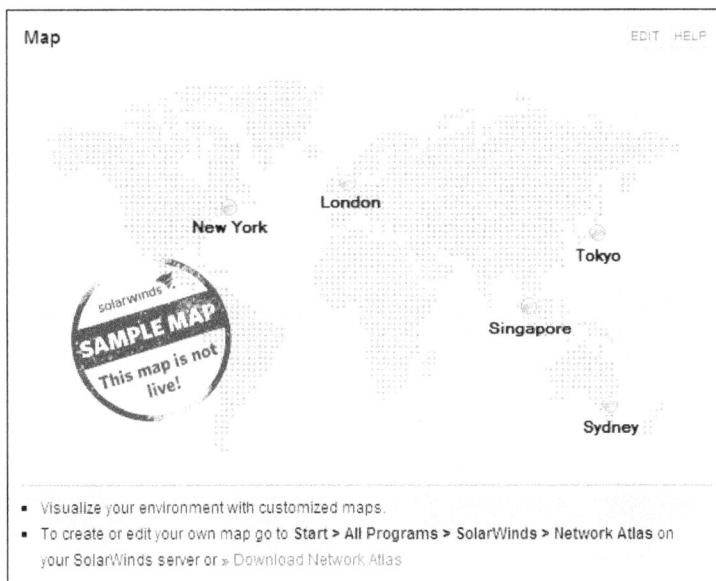

All Nodes and All Groups

The next module on the **Home Summary** page that will strike your attention is **All Nodes**. **All Nodes** displays every node that is being monitored by Orion NPM.

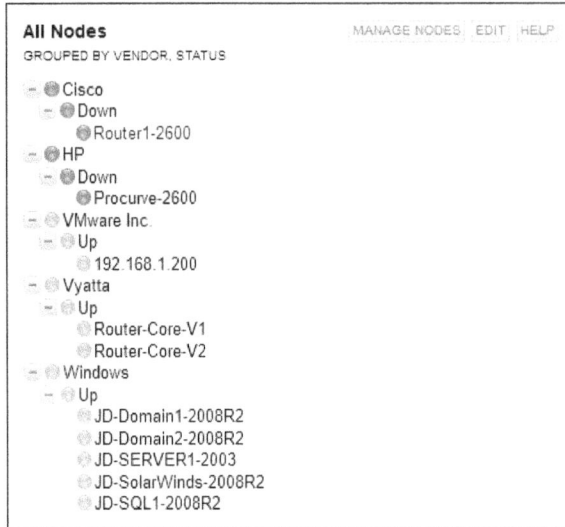

On the right-hand side of the screen underneath the network map is the **All Groups** module.

This displays any groups that I have configured in Orion NPM. This module operates in the same way that the **All Nodes** module does, aside from the fact that it shows the status of an entire custom group instead of just the nodes themselves.

The view is configurable, but by default, the module displays first by vendor, its up or down status, and then the host name. If the host name is not known, then it will display the IP address of the node. A sample of my own network lab is displayed in the previous screenshot. **All Nodes** displays what nodes are up and what nodes are not responding to Orion NPM.

All Triggered Alerts

Continuing down to the left-hand side of the page is the **All Triggered Alerts** module.

This module is very helpful in that it displays the time and date, node title, current value if available (such as **Active**), and the description of the latest alerts that Orion NPM triggered. In the preceding example, you can see that on March 11, 2013 at 8:52 A.M., Orion NPM detected high packet loss on the **JD-1130AP** node. Then one minute later Orion NPM triggered an alert that it could no longer communicate with the node. To see more details about that specific node, click on the node name to open the **Node Details View** page. We'll discuss alerts in detail in *Chapter 6, Setting Up and Creating Alerts*.

Event Summary and Last 25 Events

The next modules are **Event Summary** and **Last 25 Events**.

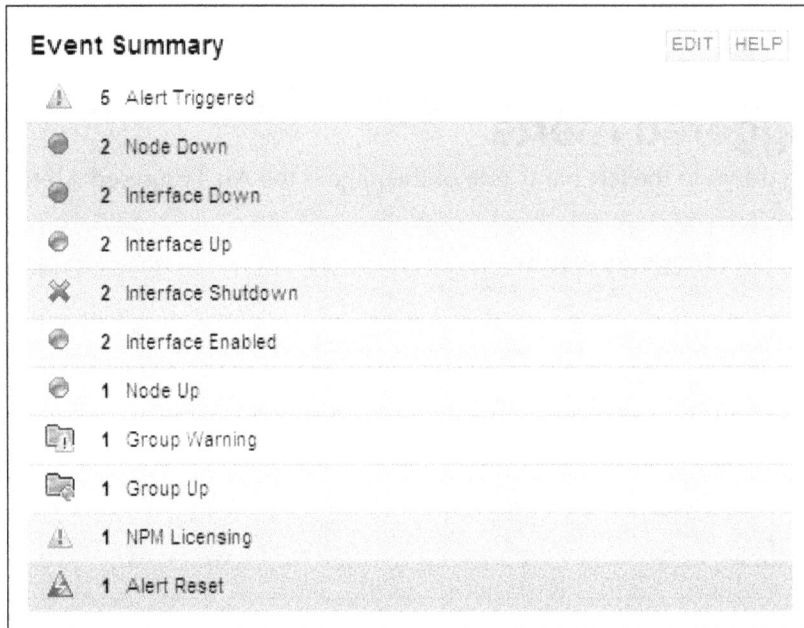

This module displays a summary list for all types of events related to network monitoring and only displays events from the last 24 hours. It is useful when needing a quick rundown of a total number of events that occurred throughout the day.

Since this is only a summary of events from the day, there is very little information provided. To see more information on a specific line of item in the **Event Summary** module, click on one of the event titles to open a filtered view in the **Events** web page. Clicking on an event title, such as **5 Alert Triggered** shown in the preceding example, will open the **Events** page with a filter to only display **Alert Triggered**. Clicking on **2 Node Down** event will open the **Events** page with a filter to only display **Node Down** information.

The final module on the **Home Summary** page is **Last 25 Events** and shows more event information in a historical context. A sample of what it displays is shown in the following screenshot:

From the example, you can see that someone caused quite a few events within a short amount of time. On the **JD-3500XL** node, you can see that a **FastEthernet** port was administratively disabled, a port labeled **Wireless Trunk** came online, the **JD-1130AP** node's packet loss rose above the loss threshold, and other important information. The downside of this module is that it will only show the last 25 events but it is extremely useful in assisting with troubleshooting a recent issue.

Search nodes

Search Nodes is a useful module where you can search a node to quickly access that node's detail view. For example, if I can't remember the name of a node but I do know where it is located, I can search the location description instead of clicking through all of the nodes until I find it. In the following example, I am searching for all nodes in the **Orlando** location.

Search Nodes		EDIT HELP
Find	Search By	
orlando	Node Name ⌄	SEARCH
Examples: Cisco*, 10.15.*.*, v	Node Name	
	IP Address	
	IP Version	
	DNS	
	Vendor	
	Description	
	Location	
	Contact	
	Status	
	IOS Image	
	IOS Version	
	CustomPollerLastStatisticsPoll	
	City	
	Comments	
	Department	

All of my nodes in the **Orlando** location are displayed in the search results. From this point, I can click on the node name to open the **Node Details View** page.

Nodes with Location similar to 'orlando':

JD-2621   Orlando, FL

JD-3500XL   Orlando, FL

Custom

Custom Object Resource is a module that allows you to create your own modules by displaying polling data of your choice. Click on the **Configure this resource** link, or the **EDIT** button, to view the contents.

Nodes with Location similar to 'orlando':

○ JD-2621   Orlando, FL

○ JD-3500XL   Orlando, FL

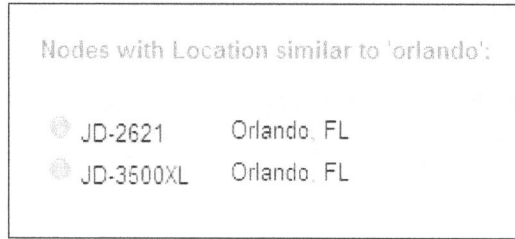

If there are some modules or resources that you do not need to view or certain pieces of information that you want to add to your pages, you can do so by customizing the web pages and modules.

Customizing views

A **view** in Orion NPM is the same thing as a web page. Views are displayed when clicking on a link in the menu bar, when clicking on a node in the dashboard, or when clicking on an interface in the nodes detail view. Each module on each page can be fully re-arranged as you see fit. It is possible that you want to view the network map on the left-hand side column instead of the right-hand side on the home summary page. You can make that change from the view editor.

Looking at the Orion Web Administration page, the **Views** module is where we are going to focus our attention.

Views
Each View can be customized. You can select which charts and device properties are displayed on each view.

» Manage Views   » Add New View   » Views by Device Type

In it are the following three links:

- **Manage Views**
- **Add New View**
- **Views by Device Type**

Manage Views

Clicking on **Manage Views** will display the **Manage Views** editor. Orion NPM sets up default views in an out-of-the-box installation. To edit a view, highlight it and click on the **Edit** button.

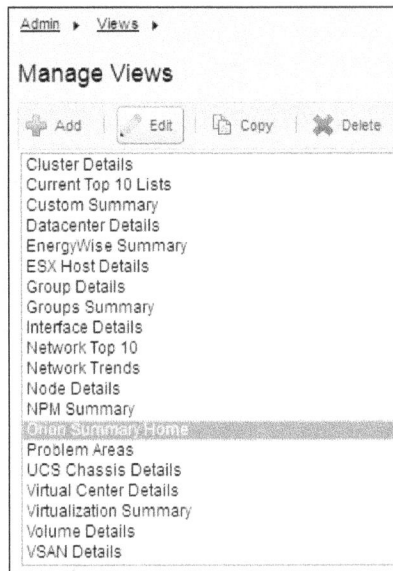

> There is no Save or Undo button when editing views in Orion NPM. Once a change has been made to a view, it is permanent. Make note of what settings are in the view you are editing before changing them in case you wish to revert back.

To demonstrate how to customize a page view, the following is an example of how to do so with the **Orion Summary Home** page. In this example I will perform the following:

- Move the **Map** from the right-hand side column to the left-hand side
- Gather **All Nodes**, **All Groups**, and **All Dependencies** together in the right-hand side column under **All Nodes**
- Place **All Triggered Alerts**, **Event Summary**, and **Last 25 Events** under the **Map** on the left-hand side column
- Remove the **Search** module
- Set the columns to be of equal width

Perform the following steps to execute these tasks:

1. Highlight **Map**, **All Triggered Alerts**, and **Event Summary** in **Column 2** then click on the left arrow button.

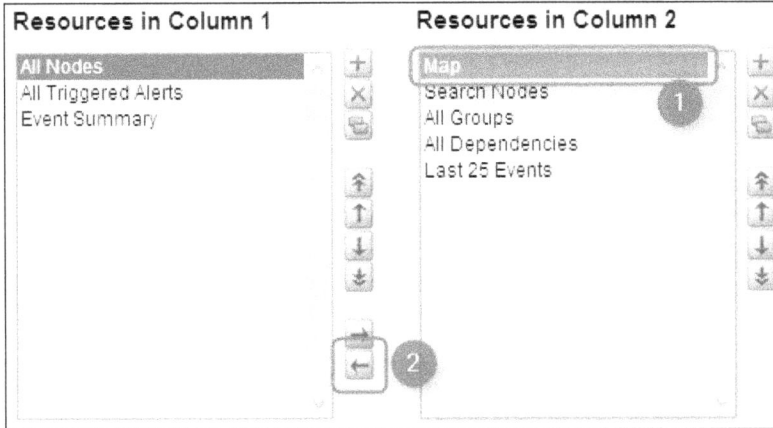

> Hold the *Ctrl* key on the keyboard in order to select multiple options in the column.

2. Reorder the **Column 1** list by using the up arrow button until **Map** is first on the list.

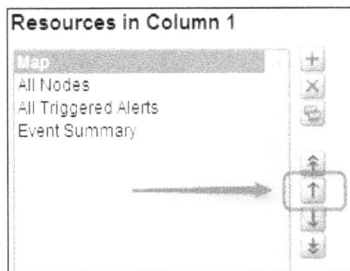

3. Highlight **All Nodes** in **Column 1**, then click on the right arrow button to move it to **Column 2**.

4. Reorder the **Column 2** list by using the up arrow button until **All Nodes** is first on the list.

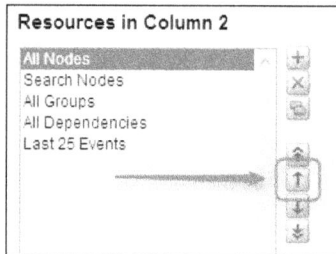

5. Highlight **Search Nodes**. Click on the red **X** to remove it from the column.

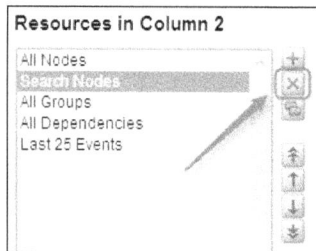

6. Click on the **Edit** button next to the column widths.

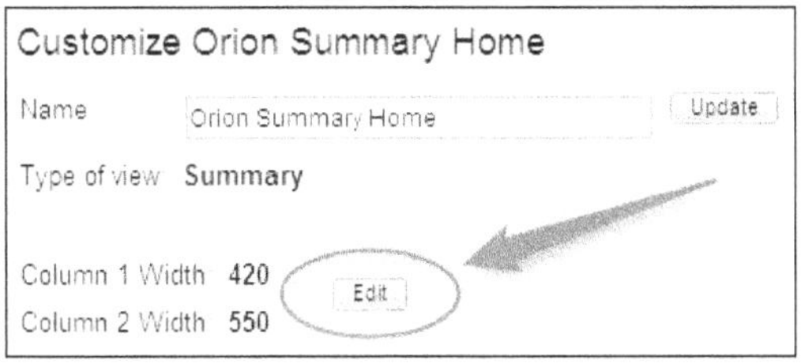

Customize Orion Summary Home

Name   Orion Summary Home   [Update]

Type of view   Summary

Column 1 Width   420   [Edit]
Column 2 Width   550

7. Change the column widths for columns **1** and **2** to **500**. Ensure that the **Layout** is set for two columns. Click on **SUBMIT** when finished.

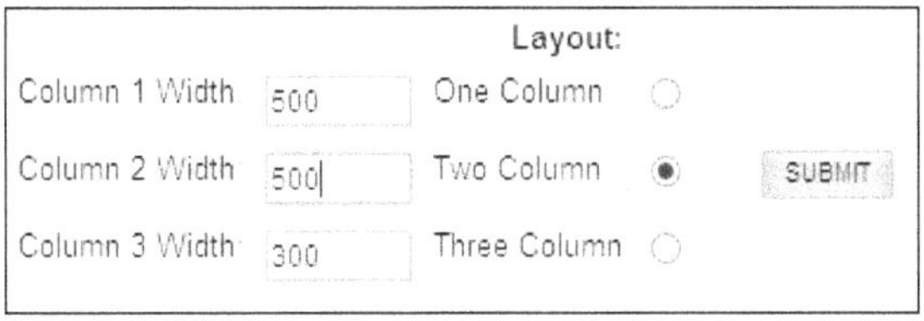

	Layout:	
Column 1 Width	500	One Column ○
Column 2 Width	500	Two Column ● SUBMIT
Column 3 Width	300	Three Column ○

Your page view should now look like the following screenshot:

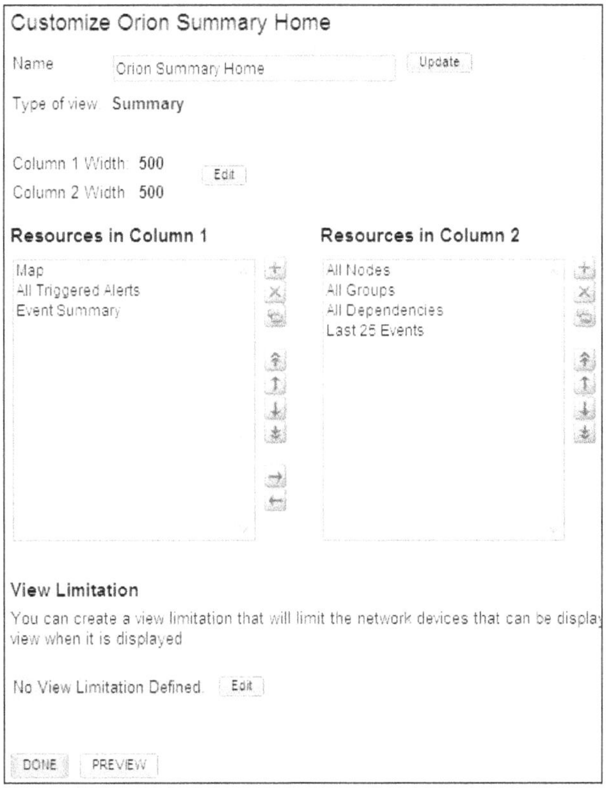

Customize Orion Summary Home

Name   Orion Summary Home   [Update]

Type of view   Summary

Column 1 Width   **500**   [Edit]
Column 2 Width   **500**

Resources in Column 1   **Resources in Column 2**

Map   All Nodes
All Triggered Alerts   All Groups
Event Summary   All Dependencies
                                Last 25 Events

View Limitation

You can create a view limitation that will limit the network devices that can be displayed view when it is displayed

No View Limitation Defined.   [Edit]

[DONE]   [PREVIEW]

The name of the view is also the web page's title. I decided not to change the name of the view for the sake of simplicity. Also, I did not apply a view limitation.

> The **Orion Summary Home** view should not have a view limitation applied since it is the main view page to Orion NPM. Applying a view limitation may omit important node information, or the limitation may render the page useless.

Click on the **PREVIEW** button to preview the view layout in a new web browser window. It will look similar to the following screenshot:

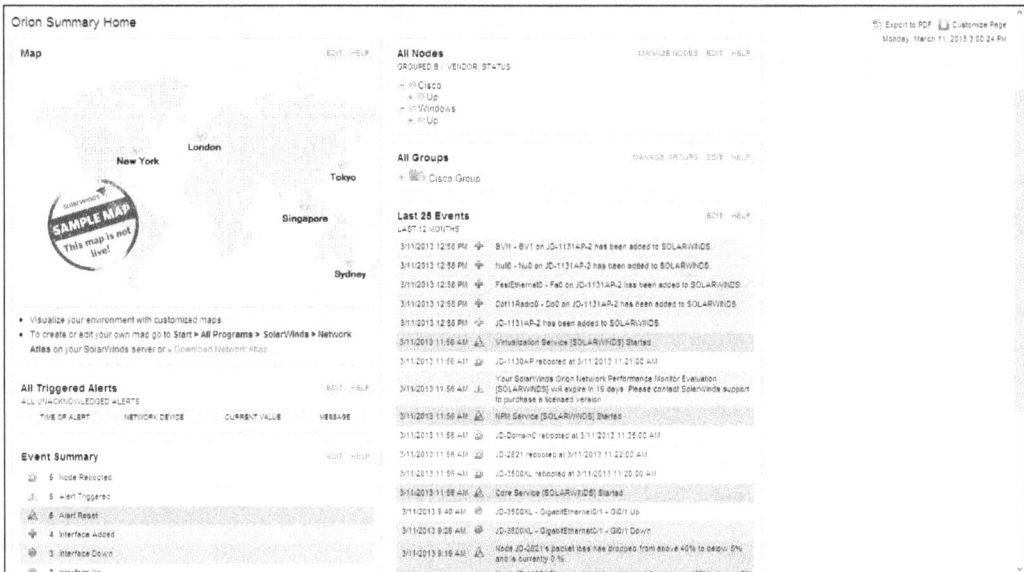

Since I was satisfied with my changes, I clicked **DONE**. As you can see, the options available when editing a view are extremely straightforward. You can change the resources (or modules) for each column, edit the column sizes, and attach view limitations. The only item you cannot change when customizing a view is the type.

> When you change the name of a view, only the title is changed. Its contents will not change.

Add New View

Orion NPM already has a great deal of default pages and default views associated with each page. However, there may be cases where a default view will not suffice and you want to create your own. The following are a few examples:

- Create a view that consists of a specific customer's equipment
- Create a view for all volumes in a VSAN
- Create a custom view for all monitored UPS units
- Create a custom view that lists all nodes in a single location

The list could go on, but you can see that there are several reasons to create your own view.

Defining the name of the view is the first step. The name you define will be the title of the page when it is saved. Make sure it is a meaningful name and one that makes sense for what you are creating.

Second, you need to choose what type of view it will be. There are several different types of views and all suit different purposes. The list of view types is as follows:

- **Summary**: Displays network-wide information. Summary is the default option. If you will be creating a view that includes multiple hardware types or locations, the Summary type will be your best option.
- **Node Details**: Displays information about a single node. This is the option that you would choose when creating a customized view page for a hardware device type. For example, you could create a custom view specifically for firewalls.
- **Volume Details**: Displays information about a single volume. Depending on your needs, you may want to create a custom view for a volume within one of your servers.
- **Group Details**: Displays information about groups.
- **Interface Details**: Displays information about a single interface.
- **VSAN Details**: Displays information about a single virtual storage area network device.
- **UCS Chassis Details**: Displays information about a Cisco Unified Computing System chassis. If you needed to create a customized view for a UCS node, this is the view type to choose.
- **Virtualization Summary**: Displays information about your VMware infrastructure.

- **Cluster Details**: Displays information about a VMware cluster.
- **Datacenter Details**: Displays information about a VMware Datacenter.

Third, you need to select the resources for each column. These are the exact same resources available when editing a view as discussed in the *Manage Views* section.

The last item that you can define is a view limitation. A **view limitation** is optional and it will limit the network devices that can be displayed within this view. An example of needing to apply a view limitation would be to limit this page to only display nodes from a specific hardware manufacturer. Or, you could add a limitation to only display nodes that reside in a specific location. The reasons why you would want to apply a limitation are virtually endless. Just keep in mind that view limitations are optional and are not required in order to create a new view.

> Only one limitation can be applied to a view. It is not possible to apply multiple view limitations.

The following is an example of how to create a custom view page with a limitation applied to only display access points:

1. In the **Add New View** wizard, enter the name `Access Points` and choose **Summary** as the view type. Click on **SUBMIT** to continue.

Add New View

Name of New View `Access Points`

Type of View `Summary`

SUBMIT

---

2. Click on the **Edit** button next to the column widths. Select the two columns and set the widths of columns **1** and **2** to `500` and `400` respectively.

Column 1 Width: 500   Edit

Column 2 Width: 400

---

3. Add resources to **Column 1** by clicking on the plus button.

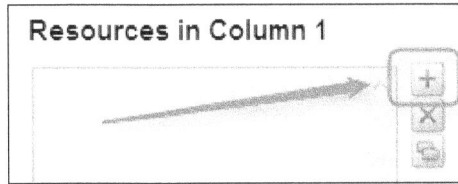

4. The **Add Resources** page appears. Expand **Node Lists – All Nodes and Grouped Node Lists** and place a check mark next to **All Nodes**. Click on **SUBMIT** to continue.

5. Add resources to **Column 2** by clicking on the plus button.

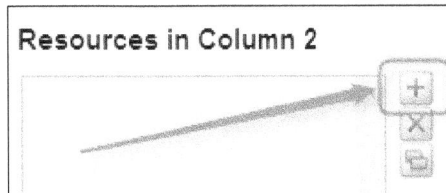

6. Expand **Summary Reports – Various Reports Showing Problem Areas** and place a check mark next to **Current Traffic on All Interfaces**. Click on **SUBMIT** to continue.

Add Resources to Access Points Column 2

⊞ Node Lists - All Nodes and Grouped Node Lists
⊟ Summary Reports - Various Reports Showing Problem Areas
    ☑ Current Traffic on All Interfaces

SUBMIT

---

7. Scroll down to **View Limitation** and click on the **Edit** button.

View Limitation

You can create a view limitation that will limit the network devices that can be displayed on this view. Account limitations for the logged-in account will also be applied to this view when it is displayed

No View Limitation Defined   Edit

---

8. Select **Group of Nodes**, scroll to the bottom of the page, and click on **CONTINUE**.

Select Limitation Type For View "Access Points"

Select a Limitation Type for this View. A Limitation can be used to limit the Network Devices that can be displayed on this view. For example, you could create a Limitation for an View that allows only Windows Servers. Or, you could limit this View to only show a specific Region or Country.

○ **Single Network Node**   Limit the Account to a Single Network Node
◉ **Group of Nodes**   Limit the Account to a group of selected Nodes

---

9. Place a check mark next to each node. In this example, the hostnames **JD-1130AP-1** and **JD-1131AP-2** are access points I am applying this new view against. Click on **SUBMIT** to apply the limitation.

10. Click on **DONE** to finish.

The following is how the new custom view will look like based on the example provided.

This is only one way to create a custom view for **Access Points**, and there are plenty of other resources that I could have added to the page. Also, I could have chosen a **Node Detail** view type instead of **Summary** type. This is only one simple example of how to create a new view in Orion NPM. I encourage you to experiment with creating your own custom views to become more familiar with the process.

Views by Device Type

Views By Device Type is where you can customize which page displays when looking at a specific type of node. For example, I can force Orion NPM to display a different **Node Details View** page for a specific model of hardware. This is helpful for when Orion NPM does not have a details view page for a hardware type, such as a UPS unit that can be monitored via SNMP. Only device types that are currently being monitored by Orion NPM will be displayed when editing views by device type.

> You cannot apply custom views against a specific hardware type unless Orion NPM is currently monitoring that type of device.

Admin ▸ Views ▸

Views by Device Type

Nodes

Object Type	Select a Web View
Cisco 2620	(default)
ProCurve Switch 2600-8 PWR	(default)
VMware ESX Server	ESX Host Details
Vyatta	(default)
Windows 2003 Server	(default)
Windows 2008 R2 Server	(default)

Interfaces

Object Type	Select a Web View
Ethernet	(default)

SUBMIT

Most options have the **(default)** option selected. The default view for almost every node monitored by Orion NPM is the **Node Details View**. Some exceptions to this rule are VMware nodes where **ESX Host Details** is automatically chosen.

Menu bars

When working with the Orion dashboard, you have already seen the tabs and menu bars at the top of the page. All of the menu bar types can be customized as you see fit. You can even create your own custom menu bars. To customize menu bars, click on the **Customize Menu Bars** link in Orion Web Administration.

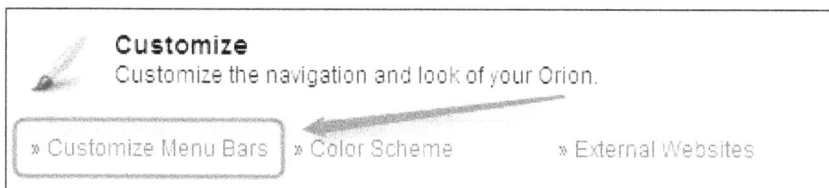

Customize
Customize the navigation and look of your Orion.

» Customize Menu Bars   » Color Scheme   » External Websites

Menu bars are assigned to one of the three tabs at the top of the dashboard. As shown in the following screenshot, there are five different menu bars from an out-of-the-box installation:

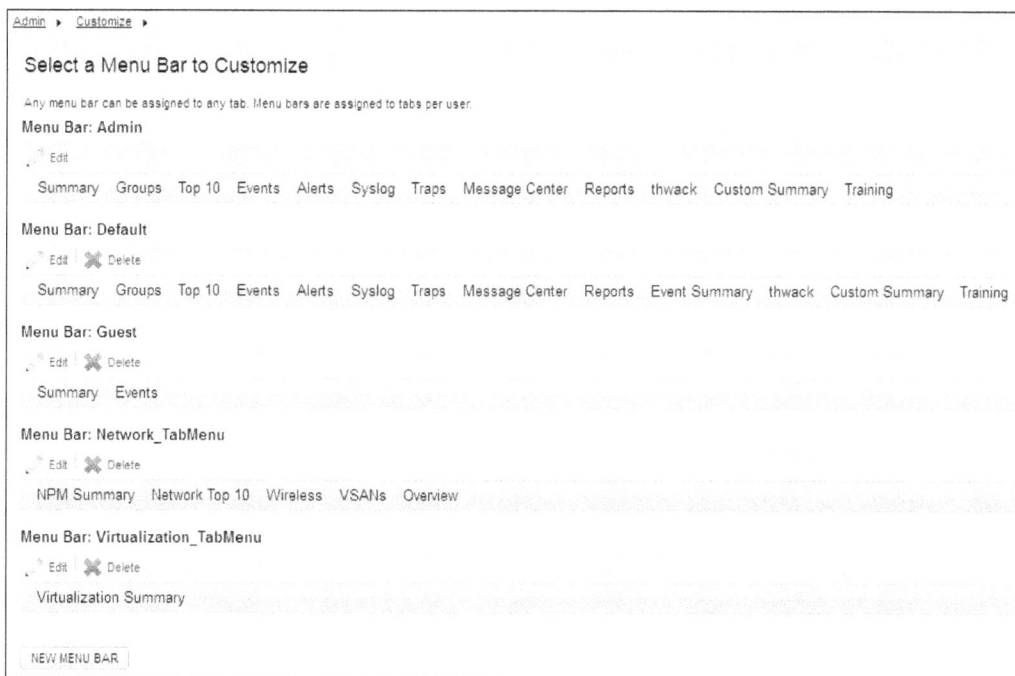

Admin › Customize ›

Select a Menu Bar to Customize

Any menu bar can be assigned to any tab. Menu bars are assigned to tabs per user.

Menu Bar: Admin

Edit

Summary   Groups   Top 10   Events   Alerts   Syslog   Traps   Message Center   Reports   thwack   Custom Summary   Training

Menu Bar: Default

Edit   Delete

Summary   Groups   Top 10   Events   Alerts   Syslog   Traps   Message Center   Reports   Event Summary   thwack   Custom Summary   Training

Menu Bar: Guest

Edit   Delete

Summary   Events

Menu Bar: Network_TabMenu

Edit   Delete

NPM Summary   Network Top 10   Wireless   VSANs   Overview

Menu Bar: Virtualization_TabMenu

Edit   Delete

Virtualization Summary

NEW MENU BAR

Orion NPM includes five default menu bars: **Admin, Default, Guest, Network_TabMenu**, and **Virtualization_TabMenu. Admin** is the only menu bar that cannot be deleted from Orion NPM but it can be edited. To edit a menu bar, click on the **Edit** button under its title and the **Edit Menu Bar** wizard will be displayed. Simply drag-and-drop the available item you wish to add to the menu bar from the right-hand side to the left-hand side column. When finished, click on the **SUBMIT** button. When creating a brand new menu bar, the same editing process applies.

In addition, menu bars are assigned to a user account's view settings through the **Manage Users** wizard. This means that when you create a new menu bar, you will need to assign it to a user account from the **Manage User Accounts** wizard.

> You cannot create your own tabs (that is Home, Network, Virtualization) in Orion NPM. You can only edit and create menu bars and assign them to a tab.

Editing Resources

While Orion NPM's default views will suit almost every need out-of-the-box, it is still a great idea to dive into all of the view settings of a module and view what you are able to customize. Every module in the dashboard allows an administrator to edit the module in some way by clicking on the **EDIT** button on the top-right corner of the resource.

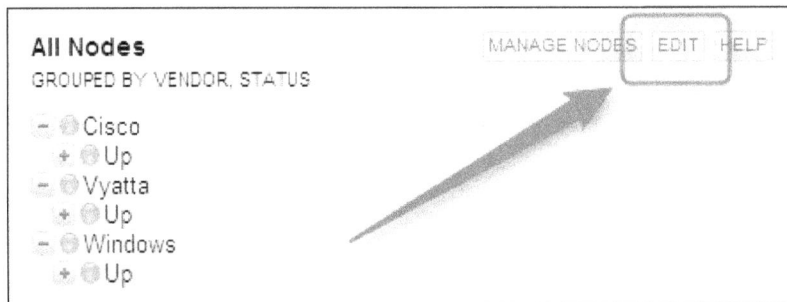

As an example, open the **All Nodes** view in the home summary screen, then click on the **EDIT** button to be presented with a list of options.

Edit Resource: All Nodes

Title:
All Nodes

Subtitle:
Grouped by Vendor, Status
Leave the subtitle field blank to have "Grouped by   " generated automatically

Grouping Nodes
Select up to three levels of grouping

Level 1
Vendor

Level 2
Status

Level 3
None

Level 1 Default: Vendor
Level 2 Default: Status
Level 3 Default: None

Put nodes with null values for the grouping property:
● In the [Unknown] group
○ At the bottom of the list, in no group

☑ Remember Expanded Groups

Filter Nodes (SQL)

Filters are optional and can be used to limit the list of Nodes displayed
This is an advanced feature. We recommend you have a basic understanding of SQL Queries

+ Show Filter Examples

SUBMIT

Every single module in Orion NPM can be edited, to a certain limit. You can always edit the title of the module as well as the subtitle in case the default descriptions are difficult to understand. For **All Nodes,** you can edit the grouping list for up to three levels. An example of creating a view for geographic locations is to set the first level to `City`, then leaving the second and third levels set to `None`. This will display only the city name at the top level, then the node names underneath. In the following example, it is easy to see how simple this type of view can be:

All Nodes   MANAGE NODES EDIT HELP
GROUPED BY CITY, THEN NODE NAME

–   Chicago, IL
        Chicago-Router1
        JD-1131AP-2
        JD-2621
–   Orlando, FL
        JD-1130AP-1
        JD-3500XL
        JD-DomainC

For medium to large network sizes, a more appropriate view option is to set the first level to Location then level two to Department. Feel free to set the grouping display to one that will suit your needs.

The remaining settings in the **Edit Resource** page are:

- **Put nodes with null values for the grouping property**
- **Remember Expanded Groups**
- **Filter Nodes (SQL)**

When Orion NPM does not know a specific property for a node, such as its location or department, the **Put nodes with null values for the grouping property** setting tells Orion NPM how to group these nodes. There are two options available. We can place nodes **In the [Unknown] group** or **At the bottom of the list, in no group**.

Placing nodes in the [Unknown] group will have Orion NPM display these nodes with unknown properties (or *blank* properties) with the group title [Unknown], which will be displayed at the top. The following is an example:

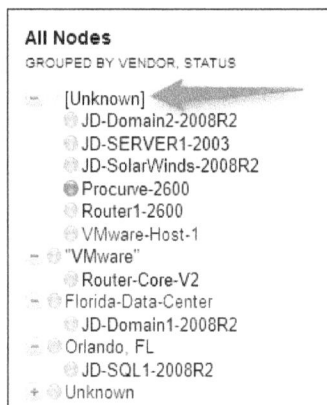

```
All Nodes
GROUPED BY VENDOR, STATUS

   [Unknown]  ◀━━━━━
       JD-Domain2-2008R2
       JD-SERVER1-2003
       JD-SolarWinds-2008R2
       Procurve-2600
       Router1-2600
       VMware-Host-1
   "VMware"
       Router-Core-V2
   Florida-Data-Center
       JD-Domain1-2008R2
   Orlando, FL
       JD-SQL1-2008R2
+  Unknown
```

The second option **At the bottom of the list, in no group** will do just that. Any node that has unknown or blank values will be placed at the bottom of the node list in the generic **Unknown** group.

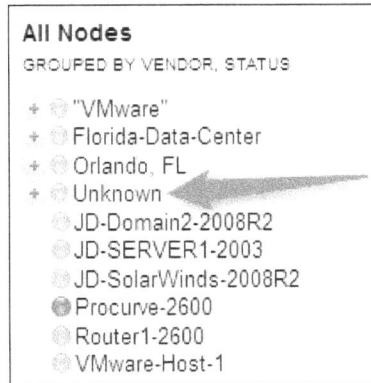

All Nodes

GROUPED BY VENDOR, STATUS

+ ⊗ "VMware"
+ ⊗ Florida-Data-Center
+ ⊗ Orlando, FL
+ ⊗ Unknown
    ⊗ JD-Domain2-2008R2
    ⊗ JD-SERVER1-2003
    ⊗ JD-SolarWinds-2008R2
    ● Procurve-2600
    ⊗ Router1-2600
    ⊗ VMware-Host-1

By default, the Orion dashboard website will trigger a browser page refresh every few minutes. When the page refreshes, if you a expanded a view in a module (a.k.a drill-down view by clicking on the plus button) the drill-down view will reset. The checkbox for **Remember Expanded Groups** is enabled by default and it is a good idea to leave this checked.

The final option is **Filter Nodes (SQL)**. This is an advanced feature of Orion NPM where you can use an explicit SQL string as a filter for these views. For example, use the filter Status<>1 to filter out all nodes that are operationally up and only view nodes that are down in the **All Nodes** module. SQL filters are helpful when creating custom views for administrative personnel. For more SQL filter examples, expand **Show Filter Examples**. Also, you can click on the **Help** button in the module for more examples and guidance on how to perform SQL queries.

> Just as in creating new views, you may have noticed by now that there is no *cancel* or *revert* option when changing a view setting in a module or a page. If you made a setting change but do not want to save the new setting, simply click on the **Back** button in your web browser to go back to the previous page without saving the new settings. Make sure that you don't change a view setting that you didn't intend to.

Customization

Orion NPM allows administrators to change a few aspects of the dashboard interface from the **Customize** module in Orion Web Administration.

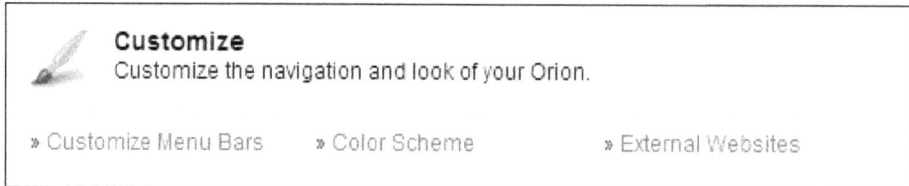

Customize
Customize the navigation and look of your Orion.

» Customize Menu Bars   » Color Scheme   » External Websites

There are three different customization options available; **Customize Menu Bars**, **Color Scheme**, and **External Websites**.

Color Scheme

Orion NPM includes several color schemes that can be changed on the fly. To change the dashboard color scheme, click on the **Color Scheme** link in Orion Web Administration, choose the **Color Scheme** option, and click on the **SUBMIT** button. Personally, I always use the **Orion Default** (white) because I can never decide which color to use!

Admin ▸   Customize ▸

Color Scheme

○ Aquamarine
○ Army Green
○ Autumn Browns
○ Firebrick
○ Lemon
○ Lilacs
○ Light Blues
○ Light Green
○ Midnight Blues
○ Pumpkin
○ Red Roses
● Orion Default
○ Steel Blues
○ Violet

SUBMIT

External Websites

The **External Websites** option in the **Customize** module is an interesting one. This option enables an administrator to add some external website to the Orion NPM dashboard as if it is a part of the dashboard itself. For example, if you have an internal Microsoft SharePoint team site on your domain, you could add it to the **Admin** menu bar and have the team site act as if it is a part of the console. When adding an external website, it must be in URL format such as `https://URL`. The following is an example of how to add an external website:

1. Click on **External Websites** in Orion Web Administration and then click on the **ADD** button.

2. Enter the **Menu Title**, **Page Title**, **URL** of the website, and which **Menu Bar** you want to apply the link to. In the following example, I am adding a link to `www.SolarWinds.com` on the **NETWORK** tab at the top of the dashboard page. Click on **OK** to finish.

The external web link will appear now appear in the **NETWORK** tab. When clicking on the link, the web page will appear as if it is embedded in the Orion dashboard.

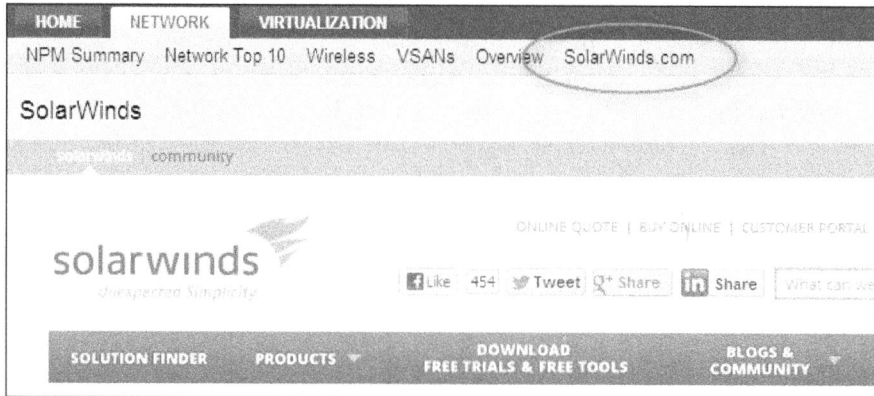

This sums up the discussion on customizing web views, modules, and other aspects of the dashboard. Now, we will discuss how to use Orion NPM to monitor your devices.

Routers and Firewalls

Knowledge of how traffic is flowing in and out of your routers is critical to maintaining network performance throughout your organization. Verifying sufficient throughput for your users is the entire point of monitoring routers and firewalls with Orion NPM. The SNMP agent installed on your devices send all types of statistical information to Orion NPM. When monitoring routers and firewalls, the items that you should be most interested in are their actual interfaces.

The default modules when monitoring routers in the dashboard include the following:

- **Multiple Object Chart**
- **Average Response Time & Packet Loss (Radial Display)**
- **Average CPU Load & Memory Utilization**
- **Node Details**
- **Average Response Time & Packet Loss (Bar Chart)**
- **Min\Max\Average of Average CPU Load**
- **Event Summary**
- **Polling Details**

- **Availability Statistics**
- **Custom Properties for Nodes**
- **Current Percent Utilization of Each Interface**
- **Virtual Machine Details**
- **Disk Volumes**
- **Active Alerts on this Node**

There is one module that is not displayed on the **Node Details View** by default, and that is the **Network Latency & Packet Loss**.

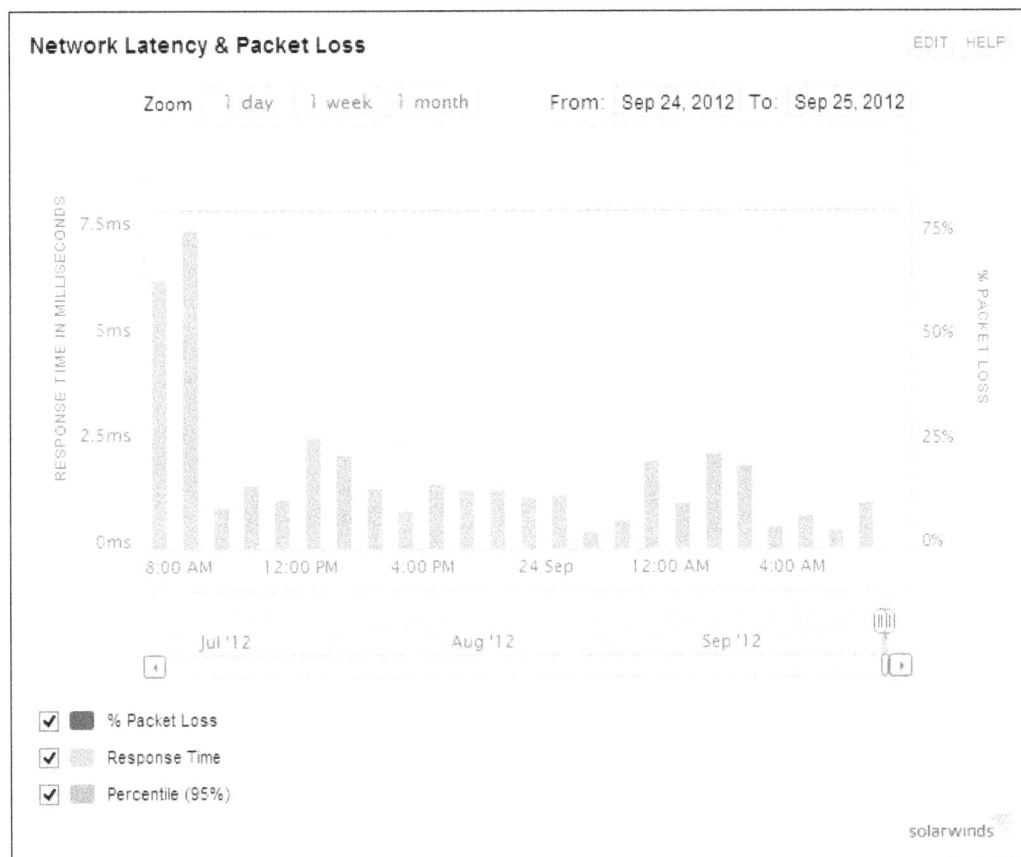

This resource is very similar to the **Average Response Time & Packet Loss** module. But there is one key difference; it can display recorded data over an extended period of time (up to several months). I recommend removing the **Average Response Time & Packet Loss** module from the **Node Details View** and adding the **Network Latency & Packet Loss** resource instead.

When monitoring routers and firewalls, the items that you want to keep an eye on the most are as follows:

- Response Time
- Packet Loss
- Throughput/Utilization
- Interface Monitoring

By default, **Average Response Time and Packet Loss** statistics for devices are displayed in two different resources on the **Node Details View** page. The first is a radial dial and the second is a bar chart. The radial dial displays captured data from the last time the device was polled and are statistics for the device as a whole, not for particular interfaces. These are the statistics that will clearly indicate if there is a network performance issue. The radial gauges displays the current response time and any recorded packet loss for this interface at this moment in time.

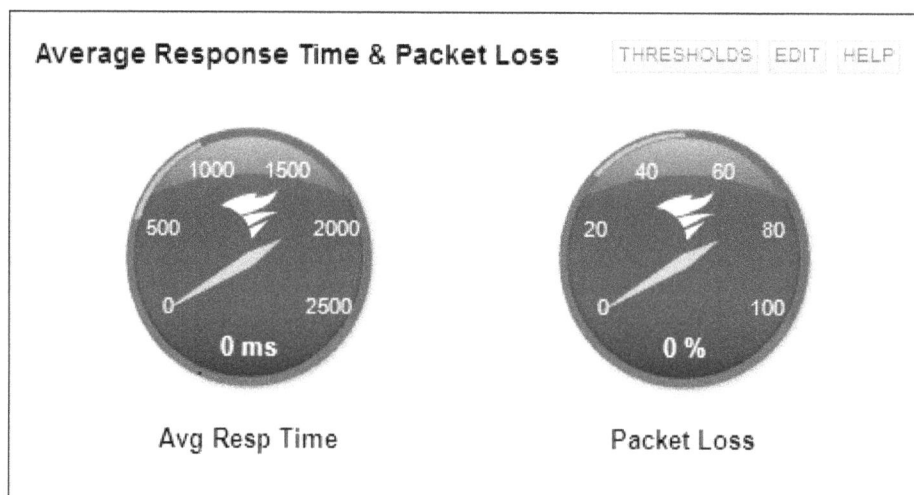

The bar chart displays the response time and packet loss over a period of time. In the following example, you can see that there are some very short instances where the response time spikes up, but goes down quickly. It is important to note that the left-hand side of this bar chart displays the highest millisecond response and does not necessarily indicate a serious problem on your network. The default view for the bar chart is 24 hours. If you want to view a larger time period, such as 7 days or 30 days, click on the **View Options** drop down and choose an appropriate option.

Average Response Time & Packet Loss | View Options ⌄ |

TODAY

Router-Core-V1
Average Response Time & Packet Loss
Today

☐ Response Time ☐ % Packet Loss

SolarWinds Orion Core Services 2012.1

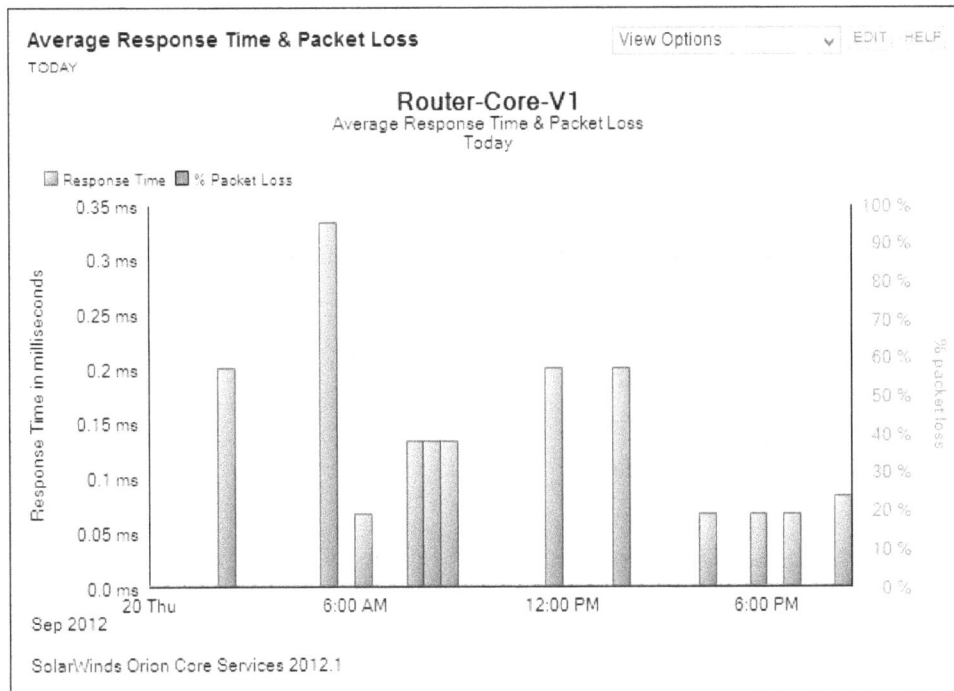

There are great deals of things that can hog up network bandwidth and network resources on your network including the basic TCP/IP overhead. Bandwidth is always limited and is an expensive commodity. So, measuring network throughput is another key aspect of router monitoring with Orion NPM. First, we do this by viewing the **Current Percent Utilization of Each Interface** resource on the **Node Details View** page.

In the following example, you can see that I have two interfaces on this specific router. The first is a WAN link to my Chicago site. The second is a WAN link to Seattle. Both **TRANSMIT** and **RECEIVE** are sitting at zero, which indicates that each interface are underutilized and basically unused. Each interface is set so that the maximum **TRANSIT** is 100 Mbps and **RECEIVE** is 1000 Mbps. To view the statistics for a specific interface, click on the interface's description to display the **Interface Details** view.

Current Percent Utilization of Each Interface   **Click to display.**

STATUS		INTERFACE	TRANSMIT	RECEIVE
⊙ Up	ᴨᴨ	eth0 · Link-to-Chicago	0 %	0 %
⊙ Up	ᴨᴨ	eth1 · Seattle-Int	0 %	0 %

The **Interface Details** page is where most of the magic happens when monitoring router interfaces. On this page are the following modules:

- **Percent Utilization – Radial Gauges**
- **Min/Max/Average bps In/Out**
- **Interface Details**
- **Percent Utilization – Line Chart**
- **In/Out Errors and Discards**
- **List of Interface Charts**
- **Total Bytes Transferred**
- **Interface Polling Details**
- **Interface Errors & Discards**
- **Maximum Traffic Today**
- **Min/Max/Average packets In/Out**
- **Event Summary for this Interface**

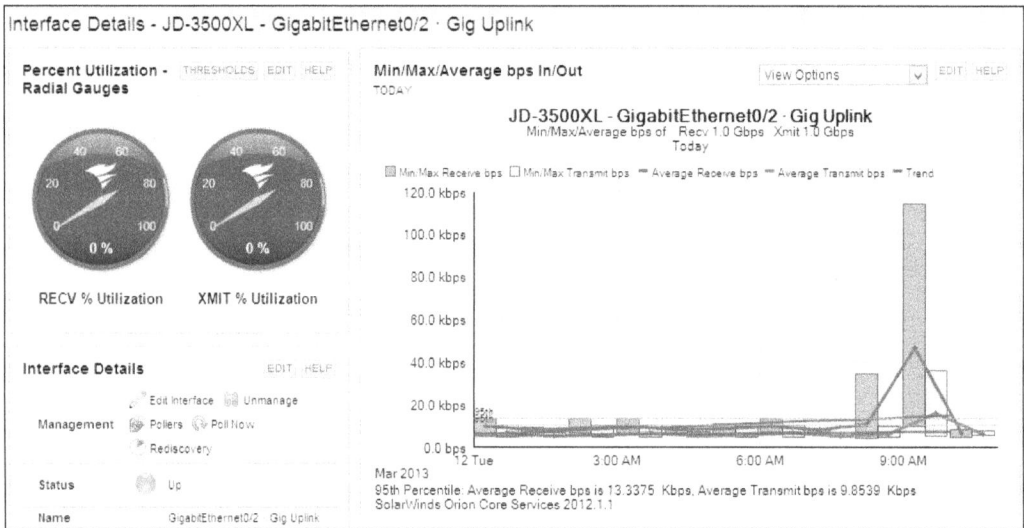

Every one of these resources is fairly self-explanatory and it is for that reason I am not going to explore every single module in detail. However, I will mention a few critical ones.

The **Min/Max/Average bps In/Out** and **Total Bytes Transferred** are the resources that will display real-time usage of the interface. These modules help with performance troubleshooting by displaying bandwidth trends over a period of time (the default of which is 24 hours). An example of the **Min/Max/Average bps In/Out** display is shown in the following screenshot:

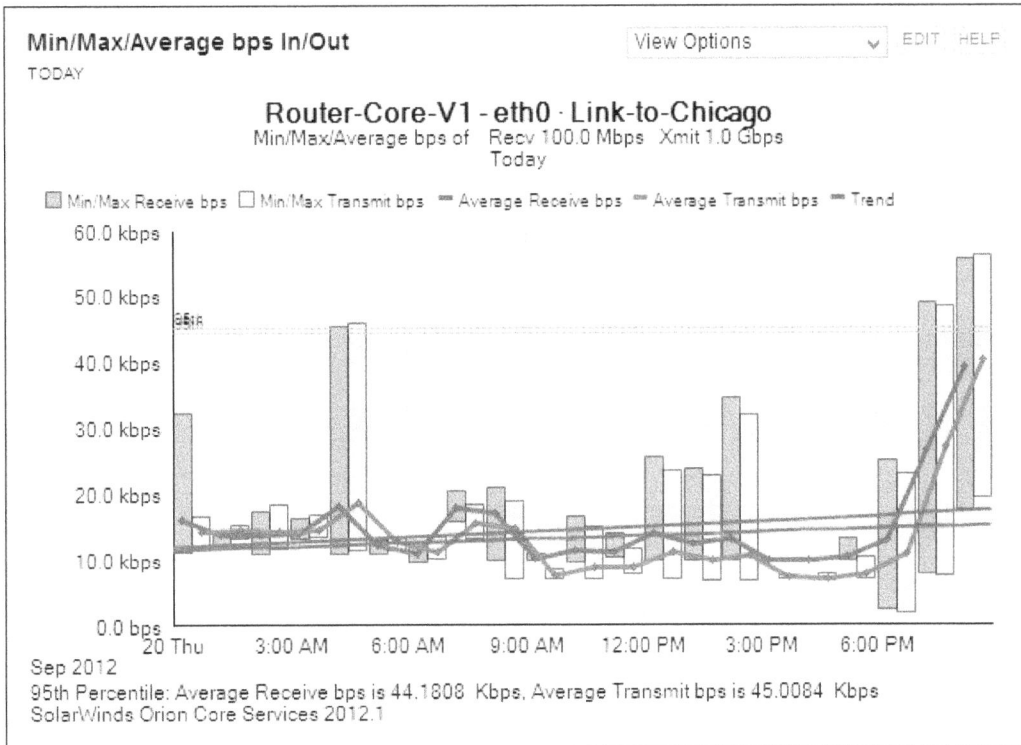

Switches

Switches are always the first point of contact for each device in your network, so if there is an issue, you want to be the first to know about it. Monitoring switches in Orion NPM is similar to monitoring routers, but with a few additional requirements.

When opening the **Node Details View** page for a switch in Orion NPM, you will find that all of the modules for monitoring a router are included here as well. You can view the response time and packet loss data, CPU and memory load, and more. However, the **All Interfaces on the Selected Node** may be your most important module when monitoring a switch. The **All Interfaces on the Selected Node** module displays all of the interfaces that are being monitored by Orion NPM. Take a look at the following screenshot:

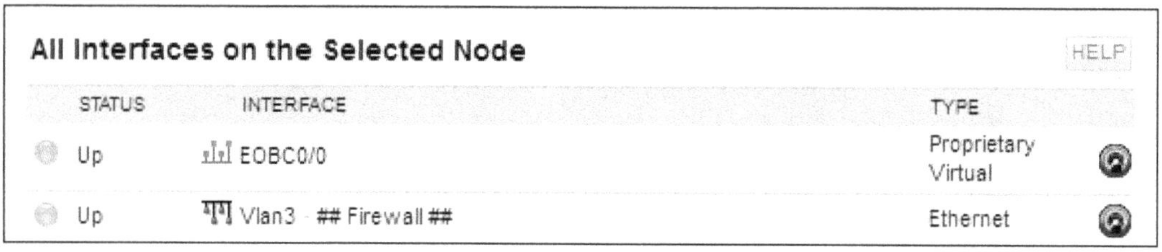

STATUS	INTERFACE	TYPE	
All Interfaces on the Selected Node			HELP
Up	EOBC0/0	Proprietary Virtual	
Up	Vlan3 ## Firewall ##	Ethernet	

There are two types of ports being monitored in this example. The first is a **Proprietary Virtual** interface type and the second is an **Ethernet** type. The first interface type is actually an EtherChannel port on the switch. Any type of virtual port will be noted in the type section of the display grid. PortChannels and Cisco StackWise ports are considered to be virtual ports in Orion NPM while VLANs and standard Ethernet ports are considered to be standard Ethernet port types. Monitoring PortChannels and stack ports is just like monitoring any other interface on a device. The same goes for monitoring VLANs.

> If Orion NPM is monitoring StackWise ports on Cisco switches, Orion NPM will not display any type of statistical data for them. This is because StackWise ports connect multiple Cisco switches as if they are one and only provide high-speed throughput for each connected switch via a virtual *backplane*. However, it is still a good idea to monitor stack ports, as Orion NPM can detect the up/down state of that port which, in effect, can indicate if one of the stacked switches is experiencing an issue.

The default modules available, when monitoring switches in the dashboard, include all of the same modules from routers and firewalls. However there is one new module applied at the bottom of the view, **NPM Network Topology**.

The **NPM Network Topology** resource is a visual representation of which interfaces from monitored nodes are physically connected to each other. It displays how they are connected by their interfaces.

As you can see in the example, the access point **JD-1130AP-1** has its **FastEthernet0** interface physically connected to the **FastEthernet0/2** interface on the **JD-3500XL** switch.

The reason why you may see an **Unknown** interface in **NPM Network Topology** is because Orion NPM must be monitoring that interface on the node. So for example, if you are monitoring a Windows Server but you are not monitoring its physical interfaces, **NPM Network Topology** will not display which interfaces are connected to the server.

Wireless

Orion NPM is a very useful application for monitoring your wireless infrastructure. It can tap into any 802.11 IEEE compliant wireless controllers or access points and provide critical information such as an access point's status, radio channels, Wi-Fi utilization, connected client details, rogue access points, signal strength and power, and more. The following screenshot displays some of the information that is captured by Orion NPM.

There are three different types of wireless devices that can be monitored with Orion NPM:

- Wireless autonomous access points
- Wireless controllers
- Wireless thin APs

Wireless autonomous access points, or thick or heavy access points, are wireless devices that can be managed and configured individually. Wireless thin APs depend on some type of centralized management device, called a wireless controller or wireless LAN controller, to retrieve its configuration. And the final type is the wireless controller itself, which is a centralized management server for thin wireless APs. Orion NPM has a view for each type of wireless node and automatically assigns the correct web view when the device has been added to the Orion database.

When a wireless controller is added to the Orion NPM database, all of the access points that are registered to that controller will be imported as well. Controllers and access points will be linked to each other in the **Wireless** summary view only after topology and interface data has been collected by the poller. For example, imagine that you have a Cisco wireless controller with five access points associated to it. You need to add the controller and all of the access points to the database before Orion NPM will poll each node.

Orion NPM does a pretty good job of auto-detecting wireless nodes and assigning them the correct web views in the dashboard. However, it is possible that an incorrect web view will be assigned to the node. If this happens, you will need to correct the web view options for that node. To do this, simply change the default view for that device type in Orion Web Administration. Follow the next set of instructions to change a web view per device type:

1. Open Orion Web Administration.

2. Under **Views**, click on **Views by Device Type**.

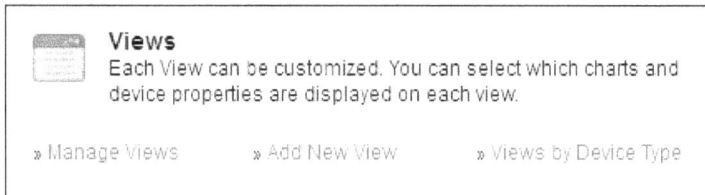

3. All of the various device types that have been detected by Orion NPM will be displayed. Choose a web view for the object type you wish to change. In the following example, I have changed the **Meraki Networks, Inc.** object type to display the **Wireless Autonomous AP** web view:

4. Click on **Submit** to save your settings.

   Once wireless nodes have been added to the Orion NPM database, the polled data can be viewed in the **Wireless Summary View** from the **NETWORK** tab in the dashboard.

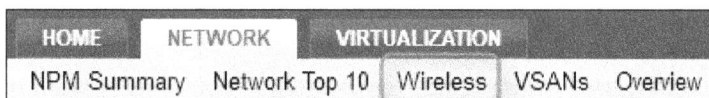

Wireless controllers and access points

Most of the details about the wireless device are shown in the display grid. Under the **Access Point** column, a red icon indicates that a node is no longer responding to the Orion pooler while a green wireless icon indicates the node is reporting back to Orion NPM without an issue. The AP's IP address, AP type, list of SSIDs, channels, and current connected client count is displayed. If there are any connected wireless clients on an access point, you can click on the carrot on the left-hand side of the access point's name. Doing this will display all currently connected clients, their MAC address, assigned IP address, signal strength, the date and time of the connection, current data rate, and inbound and outbound traffic statistics.

Access Point		IP Address	Type		SSIDs	Channels	Clients	
AustinAP1130.1		10.199.20.121	Thin	lab		48	0	
AustinAP1130.2		10.199.20.122	Thin	lab		11, 48	1	
AustinAP1130.3		10.199.20.123	Thin	lab		1, 3, 161	3	

Client Name		SSID	IP Address	MAC Address	RSSI	Connected	Data Rate	Bytes Received	Bytes Transmitted
csimmons	lab		10.199.23.6	001B53301457		10/2/2012 8:56:02 AM	54 Mbps	19.4 MB	15.7 M
emitchell	lab		10.199.23.8	001DDD556BFC		10/2/2012 9:03:01 AM	54 Mbps	240.8 MB	210.1 M
scollins	lab		10.199.23.3	0016C74B86AE		10/2/2012 8:49:17 AM	54 Mbps	35.1 MB	49.2 M

Page 1 of 1 — 10 — View 1 - 3 of

Choose the **Group By** option on the left-hand side pane to decide how you want to view the display pane of the **Wireless Summary View**. There are multiple group types available as shown in the following screenshot:

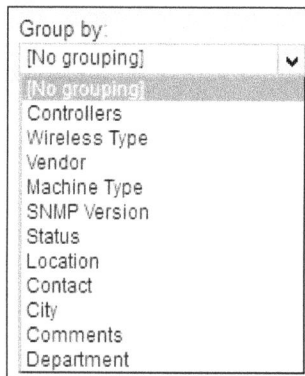

```
Group by:
[No grouping]                    ▼
[No grouping]
Controllers
Wireless Type
Vendor
Machine Type
SNMP Version
Status
Location
Contact
City
Comments
Department
```

The **No Grouping** option is usually the default display option for this view. **No Grouping** will only display access points, not wireless controllers. The only way to view wireless controllers in the **Wireless Summary View** is through the **Controllers** group.

Wireless Summary View		Show Access Points ∨			SEARCH		
Group by	Access Point	IP Address	Type		SSIDs	Channels	Clients
Controllers ∨	AustinAP1130.1	10.199.20.121	Thin	lab		48	0
Autonomous APs (2) ▸	AustinAP1130.2	10.199.20.122	Thin	lab		11 48	1
Rogue APs (23) ▸	AustinAP1130.3	10.199.20.123	Thin	lab		1 3 161	3
Aus-Cisco2106 (5)	AustinAP1130.4	10.199.20.124	Thin	lab		48	0
Bru-Aruba200 (3)	AustinAP1130.5	10.199.20.125	Thin	lab		1	0
Cai-2106 (5)							
MeruWC1 (3)							
MeruWC2 (2)							
Syd-Cisco2106 (5)							
Tok-ArubaMS200 (3)							

The **Controllers** group is helpful when needing to see which AP's are associated to which wireless controller that is being monitored by Orion NPM. The number next to each wireless controller indicates the AP count. Orion NPM classifies wireless controllers in the same category as a switch. You will find the controller in the **Network Summary** view under **All Nodes**.

Group by:
Controllers ∨

Autonomous APs (2)
Rogue APs (23)
Aus-Cisco2106 (5)
Bru-Aruba200 (3)
Cai-2106 (5)
MeruWC1 (3)
MeruWC2 (2)
Syd-Cisco2106 (5)
Tok-ArubaMS200 (3)

Another useful display group is **Wireless Type**. As stated before, Orion NPM can monitor two different types of access points, **Autonomous APs** and **Thin APs**. Thus, there are two wireless type classes for this display group. All of the other display groups are self-explanatory.

Group by:
Wireless Type ∨

Autonomous (2)
Thin (26)

The **Controllers** display group is a unique one where two very different group classes are shown; Autonomous APs and Rogue APs. Since an Autonomous AP is one that does not require a wireless controller to operate, this has its own class in the **Controllers** display group. If your wireless controller, and/or access point, has **Wireless Intrusion Protection System** (**WIPS**) features, that data will populate in the Rogue APs class. You cannot isolate rogue access points from Orion NPM. You must do that from your wireless controller.

To view the details on an access point, such as the wireless client history and interface and radio details, click on the access point in the main pane. A view similar to the **Node Details View** page will open, but only with a few extra resources that show wireless information.

The Wireless Thin AP view does not show node details such as packet loss and CPU/Memory usage, since they report back only to a wireless controller. However, Autonomous APs will show CPU and packet loss statistics because they are standalone devices. The resources that are special for both the Autonomous AP and Thin AP views in Orion NPM are as follows:

- Access Point Errors
- History Wireless Clients
- Active Wireless Clients

Access Point Errors displays any type of Energy Deferred Errors, Ack Failures, CRC Frame Errors, Transmit Errors, and Frame Check Sequence Errors per minute.

Access Point Errors	HELP
Ack Failures	0.0 / Min
Transmit Errors	0.0 / Min
FCS Errors	0.0 / Min

The **History Wireless Clients** resource displays a complete history of all wireless clients that have connected to the access point within a 24-hour period. Displayed is the IP address of the client, which SSID it is connected to, the MAC address, latest signal strength, when it got connected, and when it got disconnected. This resource will also display how much data the client transmitted and received.

History Wireless Clients

LAST 24 HOURS

	IP Address	SSID	MAC Address	Signal Strength	Connected	Disconnected	Transmitted	Received
(((p)))	10.199.25.2	lab	0010986F2621		3/11/2013 10:15:00 AM	3/12/2013 2:55:00 PM	358.0 MB	446.0 MB
(((p)))	10.199.25.7	lab	0015E840BF4C		3/11/2013 9:35:00 AM	3/12/2013 2:40:00 PM	595.9 MB	835.8 MB
(((p)))	10.199.25.4	lab	001435D31088		3/11/2013 8:40:00 AM	3/12/2013 11:40:00 AM	1.2 GB	265.5 MB
(((p)))	10.199.25.3	lab	001221A7FD64		3/11/2013 9:50:00 AM	3/12/2013 11:30:00 AM	588.4 MB	695.2 MB
(((p)))	10.199.25.18	lab	001CC2F6DE9A		3/11/2013 10:40:00 PM	3/12/2013 10:25:00 AM	2.1 GB	468.2 MB
(((p)))	10.199.25.5	lab	00157855EF4B		3/11/2013 12:10:00 PM	3/12/2013 3:45:00 AM	202.6 MB	297.3 MB
(((p)))	10.199.25.6	lab	0015834F278B		3/11/2013 11:55:00 AM	3/12/2013 2:30:00 AM	340.9 MB	372.9 MB
(((p)))	10.199.25.16	lab	0018186480B0		3/11/2013 9:15:00 AM	3/12/2013 2:25:00 AM	649.2 MB	367.3 MB
(((p)))	10.199.25.17	lab	00195C06EA53		3/11/2013 10:05:00 AM	3/12/2013 2:20:00 AM	408.4 MB	482.0 MB
(((p)))	10.199.25.1	lab	00104504B101		3/11/2013 12:30:00 PM	3/11/2013 9:40:00 PM	189.7 MB	188.5 MB

All **Active Wireless Clients** shows the current list of clients connected to the access point at this moment in time. It displays the same data as the **History Wireless Clients** resource aside from the current data rate the client has.

Active Wireless Clients

	IP Address	SSID	Signal Strength	Connected	Data Rate	Transmitted	Received
(((p)))	10.199.25.1	lab		3/12/2013 8:26:14 AM	48 Mbps	44.2 MB	53.3 MB
(((p)))	10.199.25.3	lab		3/12/2013 8:18:36 AM	54 Mbps	221.5 MB	185.1 MB
(((p)))	10.199.25.7	lab		3/12/2013 8:30:20 AM	54 Mbps	36.5 MB	50.7 MB
(((p)))	10.199.25.9	lab		3/12/2013 8:55:01 AM	54 Mbps	33.1 MB	33.0 MB
(((p)))	10.199.25.4	lab		3/12/2013 8:52:12 AM	54 Mbps	119.6 MB	72.3 MB
(((p)))	10.199.25.8	lab		3/12/2013 8:41:37 AM	11 Mbps	254.4 MB	51.6 MB
(((p)))	10.199.25.10	lab		3/12/2013 9:04:15 AM	48 Mbps	320.9 MB	53.7 MB
(((p)))	10.199.25.5	lab		3/12/2013 8:58:51 AM	54 Mbps	256.7 MB	25.7 MB

You may have noticed that the link to customize the page is missing on the top-right corner of the **Wireless Summary View** page. This is because the **Wireless Summary View** is one that cannot be edited or manipulated. You can always create your own customized view if there is some type of specific data that you want to display (such as a limited view), or you can edit the Autonomous AP and Thin AP views.

Wireless clients

The second display type for the **Wireless Summary View** is **Show Clients**.

This display type will list every currently connected device for all monitored access points in Orion NPM and it is similar to the information in the **Active Wireless Clients** resource described in the previous section. It is a simple but powerful display type due to the data that it can show about a client. This is a very helpful view when trying to troubleshoot wireless performance issues, or when trying to find out who/what/where is sucking up all of your wireless bandwidth.

Summary

In this chapter, we discussed the management and customization of web views, router and switch monitoring, and wireless monitoring. By now you can see how and why SolarWinds Orion NPM is such a popular network monitoring system and why it is loved by thousands of network and server administrators worldwide. But we are not done discussing monitoring just yet! The next chapter finishes the duology on network monitoring by discussing server, virtualization, and VSAN monitoring.

5
Network Monitoring II

And so begins the second chapter dedicated to network monitoring. The previous chapter discussed monitoring your firewalls, routers, switches, and other types of network devices. In this chapter, we are going to discuss:

- Monitoring Windows and Linux servers
- Monitoring virtual hosts running VMware ESX/ESXi
- Virtual Storage Area Networks (VSAN)
- Cisco Unified Computing System (UCS)
- Universal Device Pollers

To wrap up this chapter, we will discuss Orion NPM polling engines.

Server monitoring

Servers are the devices that provide all types of network services that are critical for business to survive. E-mail, network file sharing, FTP, and many other types of network services are what a server provides to users and customers. Monitoring servers with Orion NPM is very similar to monitoring network devices. While monitoring a server's network interface card is still available, there is a stronger focus on monitoring the CPU, memory, and disk I\O performance in servers. Thanks to Orion NPM using SNMP and WMI standards, Orion has the ability to monitor a variety of server platforms and operating systems including Windows, Linux, Mac OS X Server, Solaris, and more. I would be happy to show you how to monitor every single server OS; however, there is only so much space available in this book! Instead we will discuss server monitoring for the two most popular server platforms: Windows and Linux.

Configuring Windows servers

With Windows Server 2003, 2008, the R2 editions, and now Windows Server 2012, there is no question that the Windows Server platform is the most common server operating system in use in the enterprise environment. As you are already aware, Orion NPM operates on top of Windows Server, so it is logical that SolarWinds would provide excellent support for Windows Server monitoring. With that said, it should come as no surprise to you that Orion NPM makes it incredibly easy to monitor Windows Servers.

Before adding a Windows Server to the Orion database, you need to decide how you want to poll the node. If you recall, there are four options when wanting to add a node to the Orion NPM for monitoring:

- **No Status**
- **Status Only** (ICMP only)
- **Windows Servers** (WMI and ICMP)
- **Most Devices** (SNMP and ICMP)

If you are serious about monitoring all of the different aspects of a Windows Server such as the CPU, memory, hard disk, and interfaces, then you won't even consider selecting **No Status** and **Status Only** monitoring. In this case, you can choose either WMI and ICMP or SNMP and ICMP. But which one should you choose? It is all up to your monitoring needs. On the other hand, I'm sure you are looking for a list of guidelines, am I correct? Of course you are!

If you do not have a need to monitor interfaces in a Windows Server, then you may only need to monitor it using option number three, which is WMI and ICMP. However, if you do have a need to monitor interfaces, then you must choose the fourth option SNMP and ICMP. The reason behind this is because Orion NPM can only poll interfaces using SNMP. It is true that WMI is able to gather some types of interface data especially with the WMI SNMP Provider built into Windows Server. However, SNMP is still the universally preferred interface polling method in the industry and therefore, is what SolarWinds supports when monitoring interfaces. Orion NPM will still poll CPU, memory, and hard disk data from a Windows Server using WMI and ICMP, and SNMP and ICMP options. But if you need to monitor interfaces, you must choose SNMP and ICMP when adding the server to the Orion database.

If your Windows Server is a virtual machine in a virtual host, there is a very good chance that you do not need to monitor any interfaces on that server. This is because you are most likely monitoring the interfaces on the virtual host itself and therefore do not need to monitor the virtual interfaces on the virtual machine. On the flip side, it is highly recommended to monitor interfaces on a Hyper-V host or on a server where Hyper-V services are enabled. **Hyper-V** is a hypervisor service included in Windows Server 2008 and above which effectively turns your Windows Server into a virtual host (just like VMware or Citrix XenServer). If you are running Hyper-V services, then you most likely have multiple Ethernet interfaces installed in this server. Depending on the Hyper-V configuration, it is possible that if just one of the interfaces is down, naturally any virtual machines tied to that interface will lose network access. An excellent alternative is to completely omit monitoring interfaces on a server and just monitor ports that the server is connected to on a router or switch, provided that you are monitoring that router or switch with Orion NPM.

The first steps you must complete before monitoring a Windows Server are configuring SNMP services, setting up a user account that will have permissions to execute WMI queries, and allowing protocol port communications through the Windows Firewall.

Since you can only choose one polling type when monitoring a Windows Server, WMI and ICMP or SNMP and ICMP, you do not need to configure SNMP services on your server if you do not plan on using SNMP and you do not need to prepare a user account with the appropriate permissions if you do not plan on using WMI to monitor your server.

Configuring the Windows Firewall

Configuring Windows Firewall to allow incoming ICMP (PING), WMI, and SNMP requests can be done from the command line and is the easiest way to add firewall rules. However, depending on the version of Windows Server you are running, there are different commands to perform this task. Always execute these commands from an elevated command prompt.

ICMP

For Windows Server 2003 and 2008:

```
netsh firewall set icmpsetting 8 enable
```

For Windows Server 2008 R2 and Windows Server 2012, execute the following in an elevated command line window:

```
netsh advfirewall firewall add rule name="ALL ICMP V4
protocol=icmpv4:any,any dir=in action=allow
```

```
netsh advfirewall firewall add rule name="ICMP Allow incoming V6 echo
request" protocol="icmpv6:8,any" dir=in action=allow
```

WMI

For Windows Server 2003 and 2008:

```
netsh firewall set service RemoteAdmin enable
```

```
netsh firewall add portopening protocol=tcp port=135 name=DCOM_TCP135
```

```
netsh firewall add allowedprogram    program=%windir%\system32\wbem\
unsecapp.exe name=WMI
```

```
netsh firewall add allowedprogram    program=%windir%\system32\dllhost.exe
name=Dllhost
```

For Windows Server 2008 R2 and Windows Server 2012:

```
netsh advfirewall firewall set rule group=  "windows management
instrumentation (wmi)" new enable=yes
```

```
netsh advfirewall firewall set rule group=  "remote administration" new
enable=yes
```

SNMP

For Windows Server 2003 and 2008:

```
netsh firewall set portopening protocol = TCP port = 161
```

For Windows Server 2008 R2 and Windows Server 2012:

```
netsh advfirewall firewall add rule name="SNMP" protocol=UDP
localport=161 action=allow dir=IN
```

Configuring WMI

By default, only the `local Administrators` group has remote permissions to execute WMI queries on Windows computers. You will need to create a user account with local administrator permissions.

Configuring SNMP services

Enabling and configuring SNMP services in Windows Servers is something that still cannot be managed using Group Policies. Also, it still remains difficult to perform SNMP configuration tasks from the Windows command line. Yes, you can create an administrative script to do the job for you but the easiest way to set up SNMP on a Windows Server is manually with a mouse and a keyboard. First, an administrator must enable and install the SNMP services in the Windows Server. After the services are installed and running, SNMP can be configured. The process to do this differs between all major versions of Windows Server 2003, 2008, and 2012. SNMP is considered a *feature* in Windows Server and is enabled depending on how that version of Windows enables features. Once the SNMP services are installed, SNMP agents and Traps are configured from the Services MMC applet (`services.msc`), or from the **Administrative Tools MMC** under the **Services** section.

Enabling SNMP services in Windows Server 2003

Most services and features in Windows Server 2003 are enabled or disabled from **Add / Remove Programs** in the **Control Panel** by using **Add/Remove Windows Components**:

1. Click on **Start | Control Panel | Add / Remove Programs**.
2. Click on **Add / Remove Windows Components**.
3. Highlight **Management and Monitoring tools** and then click on the **Details** button.
4. Place a check mark next to **Simple Network Management Protocol** and click on **OK**.
5. Finish the configuration changes by clicking on **Next**. Insert the Windows Server 2003 installation disc if prompted to do so during the installation.

Enabling SNMP services in Windows Server 2008

Enabling SNMP services in Windows Server 2008 is slightly different than in Windows Server 2003 and is done from **Server Manager**:

1. Open **Server Manager**.
2. Under the server name, click on **Features** and then choose **Add Features**.
3. Place a check mark next to the **SNMP Services** option.
4. Click on **Next**, and then on **Install** to start the installation process.
5. Reboot the computer if prompted to do so.

Enabling SNMP services in Windows Server 2012

In Windows Server 2012, you must use the **Server Manager** in order to configure SNMP services. Perform the following steps to enable SNMP in Windows Server 2012:

1. Launch **Server Manager**.
2. On the left-hand side, click on **Local Server** and then on **Manage**. Choose **Add Roles and Features**.
3. At **Select Installation Type**, choose the **Role-Based or feature-based installation** and then click on **Next**.
4. In **Select Destination Server**, make sure your local server is displayed in the **Server Pool** and then click on **Next**.
5. Skip the **Server Roles** screen by clicking on **Next**.
6. Place a check mark next to **SNMP Services** and then click on the **Add Features** button to include the required features.
7. Click on **Next** to continue and then click on **Install** to start the process.
8. Reboot the server if prompted to do so after the installation has completed.

Configuring the SNMP agent in Windows Server

Once the SNMP services are installed, the SNMP agent and SNMP Traps can be configured from the Services MMC applet. The process for doing this is exactly the same on all versions of Windows Server:

1. Launch `services.msc` from the `Run` command, or open the **Services** applet from **Computer Management**.
2. Find **SNMP Service** in the view pane. Right-click and choose **Properties**.
3. In the **General** tab, verify that the service is set to **Automatic**.
4. Leave the default options in the **Log On** and **Recovery** tab as is.
5. On the **Agent** tab, enter the **Contact** and **Location** information for this server. It is recommended to leave all of the options checked under the **Service** section, especially if you want to monitor the physical interfaces on this server.
6. The **Traps** tab is where you enter your SNMP Trap settings. Add the **Community name** in the textbox and click on the **Add to List** button.
7. Choose the community name from the drop-down list and then click on the **Add** button under the **Trap destinations** section to add IP addresses and/or hosts which will be sent SNMP Trap data. When finished, click on **Apply**.

8. The **Security** tab defines what hosts are allowed to poll the server for SNMP data. The community names and specific rights the community has access to on the server must be defined in order to poll the server using SNMP.

> The **Accept SNMP packets from any host** option tells the SNMP service to allow *any* host SNMP access to this server. It is highly recommended to define precisely what hosts, by IP address, are allowed to poll for data from this server.

9. Click on **Apply**, and then on **OK** to save and close the SNMP service settings.

There is no need to configure the SNMP Trap service even after configuring the SNMP Trap information in the SNMP service. Traps are executed by Windows when it detects a problem.

That's it! Now your Windows Server(s) are configured for SNMP.

Configuring Linux servers

Now, we can't leave out Linux server monitoring can we? Of course not! Orion NPM is fully capable of monitoring not only Windows Servers, but also server operating systems based on Linux. Linux has different nomenclature regarding services and features. A daemon is the equivalent of a Windows service in Linux.

Configuring the SNMP daemon

The fastest way to set up the SNMP daemon in Linux is by using the **Terminal** (Linux's command line interface). There are some graphical user interfaces for the Linux Firewall, but this task is relatively easy to do from the Terminal.

Because there are many different variants of Linux, I am only going to demonstrate how to enable and configure the SNMP daemons on Ubuntu:

1. Launch the Terminal.

2. Install the SNMP daemon by executing the following commands in order:
   ```
   sudo apt-get install snmpd
   ```

3. Edit the SNMP daemon configuration file by executing the following command:
   ```
   sudo gedit /etc/snmp/snmpd.conf
   ```

4. Use the comments embedded throughout the `snmpd.conf` document to configure SNMP on this server. Here is an example:

```
rocommunity SolarWinds 192.168.0.224/24

rwcommunity joeSNMP JD-SolarWinds.joedissmeyer.local
```

5. Save the file and close the text editor.

6. Restart the SNMP daemon by executing the command from the Terminal window:

```
sudo /etc/init.d/snmpd restart
```

Your Ubuntu server is now configured for SNMP and can be added to the Orion NPM database.

Monitoring servers in Orion

Once you have a server added to the Orion NPM database, you can begin monitoring many aspects of that server. Server monitoring in Orion NPM is performed in the exact same way that you monitor a switch or a router, but with a few differences. When monitoring network devices such as switches, routers, or firewalls, you are focused on keeping track of interface statistics, syslog data, and the state of the device. You should still be interested in monitoring all of those same statistics on a server; however, you will want to keep a closer eye on the physical resources of the server which includes CPU, RAM, and hard disk activity. So, let's take a look at a server that I am monitoring right now for an example.

The **Node Details View** page as shown in the following screenshot, is from a Windows Server 2008 R2 server, which is functioning as a DHCP server, DNS server, and Domain Controller. Right away you may notice a difference in the detailed information view:

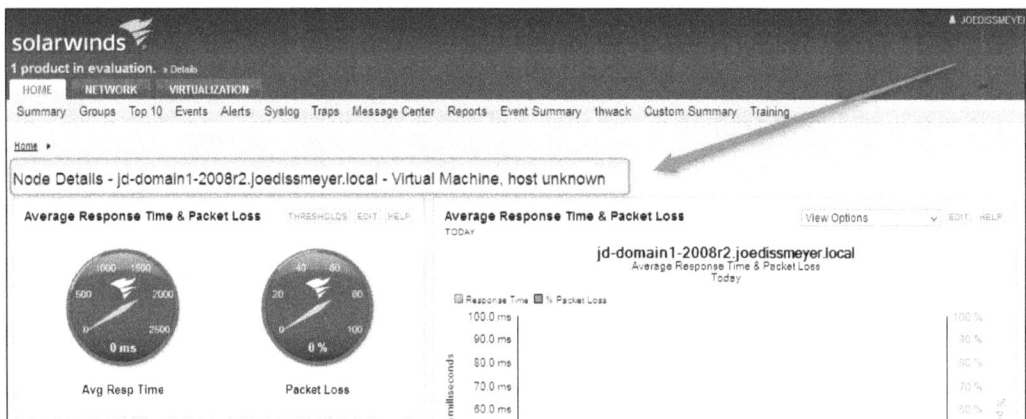

The first line displays the current view and the node name. The **Node Details View** page is the default view for all servers in Orion NPM, which is exactly the same for network appliances. You will notice that the node name is also displayed as normal, but there is something else as well. Orion NPM has detected this server as running as a **Virtual Machine**. **Host unknown** means that Orion NPM does not know exactly what virtual server is serving the virtual machine.

Moving further down the view screen, you will notice that the exact same modules are displayed for a switch or router.

- **Average Response Time & Packet Loss**
- **Average CPU Load & Memory Utilization**
- **Average Response Time & Packet Loss**

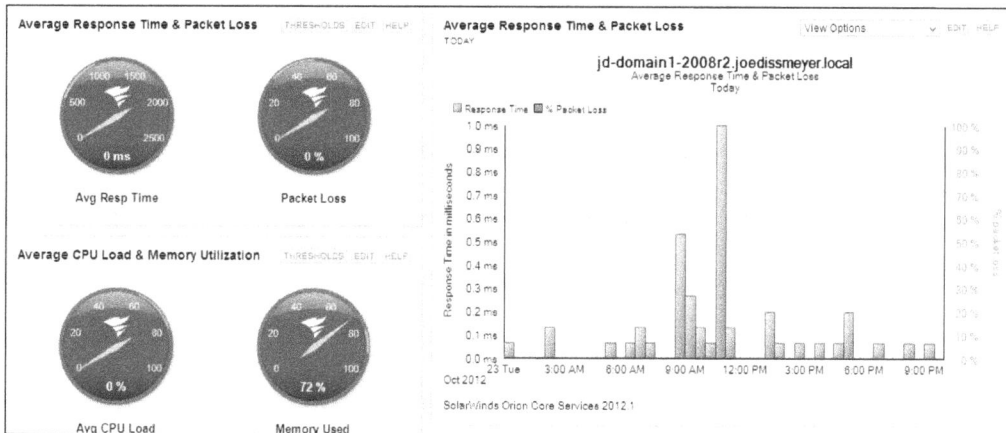

These are the three primary resources in the **Node Details View** page that informs you about the current state of the server. The radial dials and bar grid for response time and packet loss report against the network interface(s) of the server, and the radial dial for CPU load and memory utilization gives a quick look at the current server load. If you are not monitoring the server's network interfaces, then the statistics will read zero.

Moving down the details view is the **Min/Max/Average of Average CPU Load** resource. By default, the view option for this module displays data for a period of 7 days but can be modified:

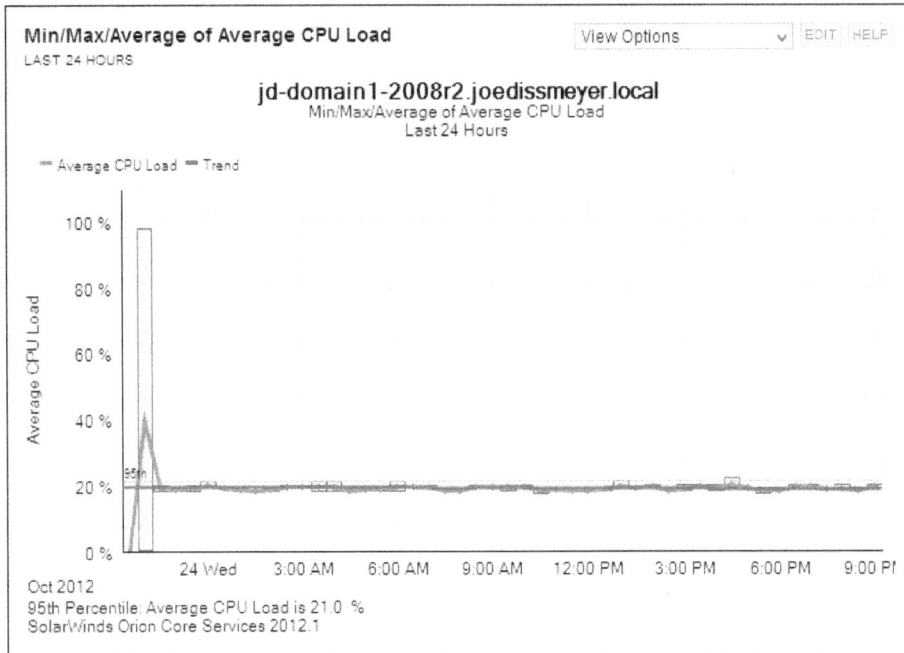

From the preceding screenshot, you can see that my server has literally been doing no work. Then all of a sudden there is a burst of activity. Being able to view this burst of activity in a specific time slot enables administrators to research an issue more quickly. This resource is extremely helpful when troubleshooting server performance issues.

The **Node Details** resource displays a great deal of information about the server including how it is being polled, its IP addressing information, what operating system is running on the server, details about the hardware, and other information. The following information is very similar to what you will see for a network device:

Node Details

EDIT HELP

Management	Edit Node List Resources Unmanage Pollers Poll Now Rediscover
Node Status	Node is Up. Interface 'WAN Miniport (Network Monitor)-QoS Packet Scheduler-0000 - Local Area Connection* 7-QoS Packet Scheduler-0000' has state: Unknown.
IP Address	192.168.1.220
Dynamic IP	No
Machine Type	Windows 2008 R2 Domain Controller
DNS	jd-domain1-2008r2.joedissmeyer.local
System Name	JD-Domain1-2008R2.joedissmeyer.local
Description	Hardware: Intel64 Family 6 Model 15 Stepping 6 AT/AT COMPATIBLE - Software: Windows Version 6.1 (Build 7601 Multiprocessor Free)
Location	Orlando, FL
Contact	Joe Dissmeyer
SysObjectID	1.3.6.1.4.1.311.1.1.3.1.3
Last Boot	Saturday, October 13, 2012 6:54 PM
Operating System	6.1 (Build 7601 Multiprocessor Free)
IOS Image	Unknown
Hardware	Virtual, host unknown
No of CPUs	1
Telnet	telnet://192.168.1.220
Web Browse	http://192.168.1.220

The last two entries in the **Node Details** screen show a **Telnet** and **Web Browse** link. These are hard coded into the resource and cannot be edited.

We have already covered the other modules in this view in a previous chapter. However, I do not want to pass up talking about the **Disk Volumes** resource:

Disk Volumes

EDIT HELP

	VOLUME	SIZE	SPACE USED		
	C:\ Label: 3A21D619	39.7 GB	13.0 GB	33 %	
	Physical Memory	1023.6 MB	558.3 MB	55 %	
	Virtual Memory	2.1 GB	673.2 MB	32 %	

The **Disk Volumes** module displays critical information about the hard disks, RAM, and virtual memory of a server. In the preceding screenshot, you can see that I am monitoring the C: of the server, the RAM—**Physical Memory**, and the **Virtual Memory**. The RAM is already at **55%** usage while the **Virtual Memory** is well below. Clicking on any of the volume titles in this module will open up the following **Volume Details** view:

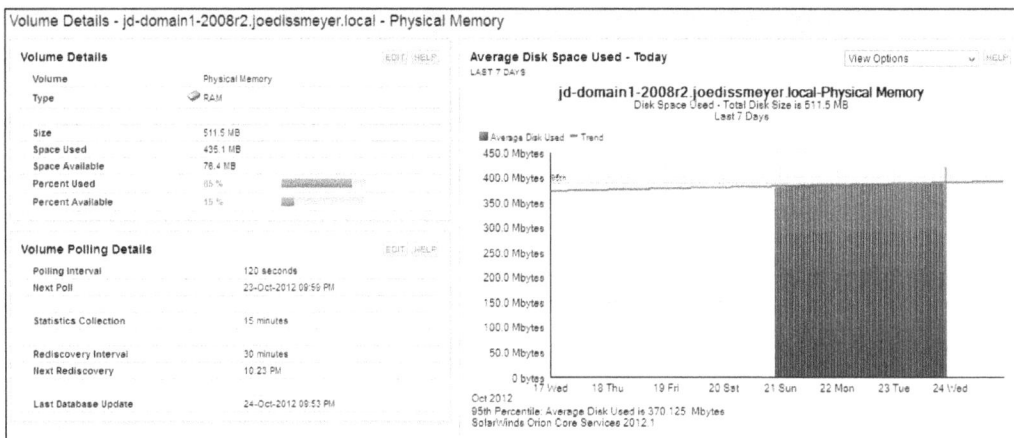

The resources listed in **Volume Details** are as follows:

- **Volume Details**: This tells you everything there is to know about the statistics for a specific disk, be it memory, virtual memory, or a hard disk.

- **Volume Polling Details**: This displays Orion NPM polling information which is inherited from the node itself.

- **Average Disk Space Used**, **Percent Disk Space Used**, and **Volume Size**: This module displays historical data about the volume's usage which is helpful for capacity planning.

Back to the **Node Details** screen, let's discuss server availability. Orion NPM keeps historical data of the up/down status of a node via the **Availability Statistics** module. Keeping high **Service Level Agreement (SLA)** percentages is a part of working in the IT field and this resource will help you with obtaining those metrics.

Availability Statistics

PERIOD	AVAILABILITY
Today	100.000 %
Yesterday	100.000 %
Last 7 Days	89.024 %
Last 30 Days	89.024 %
This Month	89.024 %
This Year	89.024 %

HELP

As you can see in the preceding screenshot, the availability of this server has been tapped out at **100%**. According to Orion NPM, this means that Orion NPM has been able to maintain a constant and consistent polling interval with the server. However, over the past few days there has been reduced availability. Clicking on one of the availability periods, such as **Last 7 days**, will display the availability metric in a new page:

Availability Statistics

PERIOD	AVAILABILITY
Today	100.000 %
Yesterday	100.000 %
Last 7 Days	89.024 %
Last 30 Days	89.024 %

Click to view

The metric display for **Last 7 days** is as follows:

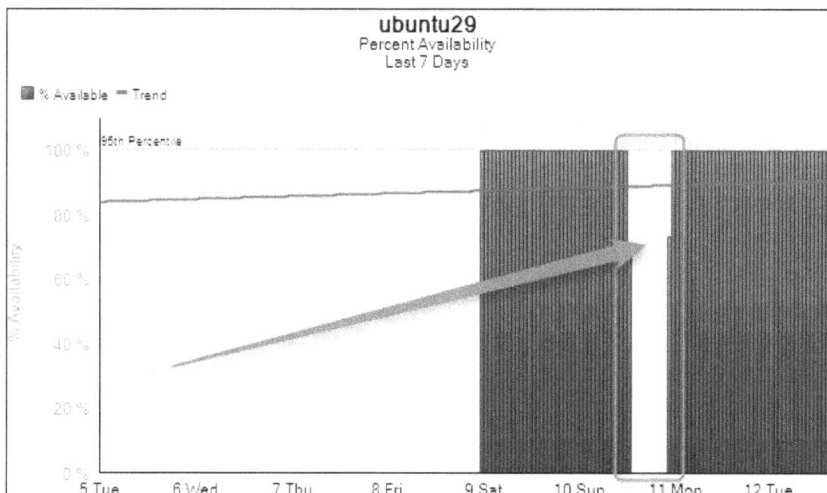

In the metric display, we can see there was some type of availability issue in between Sunday and Monday. You can now take this information and use it to find out if there were issues with other applications during this time.

Servers must be rebooted from time to time after installing updates or for other administrative reasons. With that being said, this statistic is very useful when trying to determine if a server application is causing problems, if a server needs to be moved, or if a server needs to be replaced.

Virtualization monitoring

Virtualization is one of the hot topics in IT these days. One of the things commonly lacking in a virtualization deployment is a proper way to monitor virtualization environments. Thankfully, SolarWinds Orion NPM has come to the rescue with built-in virtual host and VSAN monitoring.

> Orion NPM can only monitor VMware hosts. Citrix XenServer, Microsoft Hyper-V, and other virtual hosts are not supported.

All virtual hosts that are monitored by Orion NPM will be displayed in the **Virtualization Summary** view under the **Virtualization** tab.

Virtualization summary

Let's dive right in and start looking at a physical VMware host that is being monitored by Orion NPM right now.

Looking at the **Virtualization Summary** view, you can see all of your VMware hosts that are being monitored by Orion NPM in a single pane of glass. Think of this view like Orion NPM displaying *the big picture* of your entire virtual infrastructure.

SolarWinds uses the VMware API to poll for data from VMware hosts and VMware vCenter. Whichever way you add a virtual host to the Orion database, the end result is the same. You will want to create a separate user account in ESXi or vCenter, for SolarWinds Orion NPM to use, that has permissions to poll the virtual hosts. If you intend on monitoring interfaces on the VMware host, you will also need to configure SNMP services on the host. Remember that Orion NPM can only poll interfaces using SNMP.

> For information on how to configure SNMP on your VMware hosts refer to http://blogs.vmware.com/vsphere/2012/11/configuring-snmp-v1v2cv3-using-esxcli-5-1.html.

Assuming you have already added a VMware ESX host to the Orion NPM database, the first stop when viewing the **Virtualization Summary** view is the **VMware Assets** resource. **VMware Assets** lists all monitored virtual hosts, as well as all of the virtual machines associated with the host:

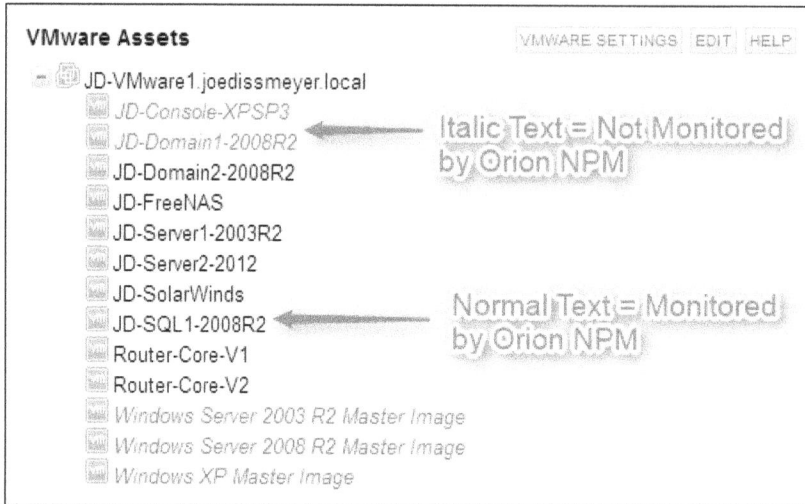

Any virtual machine that is not currently monitored by Orion NPM has its hostname displayed in italic text, while nodes that are monitored by Orion NPM are displayed normally. If you want to monitor one of those virtual machines as indicated by italic text, simply click on it and the Orion Dashboard will pop up an information window asking if you want to add it to the Orion NPM database:

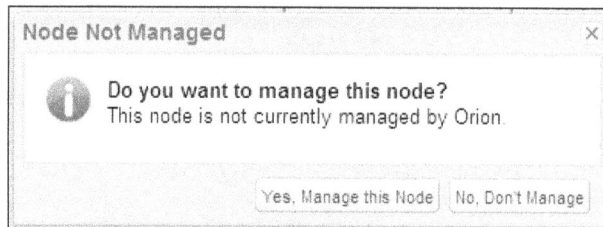

Otherwise, click on a monitored node to display the **Node Details** screen for that node.

It is possible that you do not have a need to monitor some virtual machines. But if you do need to monitor a VM, this process makes it much easier to add it to the database on-demand. Just keep in mind that for every virtual machine added to the Orion NPM database, it will use additional licensing.

The next resources in the **Virtualization Summary** view are the top 10s:

Top 10 VMware Hosts by Percent Memory Used   EDIT HELP

VMWARE HOST	MEMORY USED
JD-VMware1.joedissmeyer.local	7187 MB 89 %

Top 10 VMware Hosts by Network Utilization   EDIT HELP

VMWARE HOST	RECEIVE	TRANSMIT	TOTAL	UTILIZATION
JD-VMware1.joedissmeyer.local	0.02 Mbps	0.01 Mbps	0.03 Mbps	0 %

Top 10 VMware Hosts by CPU Load   EDIT HELP

VMWARE HOST	CPU LOAD
JD-VMware1.joedissmeyer.local	12 %

Top 10 VMware Hosts by Number Of Running VM's   EDIT HELP

VMWARE HOST	# RUNNING VMS
JD-VMware1.joedissmeyer.local	10 of 13

These are the resources that truly give a *big picture* overview of your monitored virtual hosts. When reviewing your capacity planning, or disaster recovery planning, these modules can help guide you in the right direction fast. Also, these resources help administrators quickly locate issues with capacity or while troubleshooting other issues.

The **VMware Asset Summary** view is a simple *what do I have* list concerning your virtual nodes and virtual machines. It is a complete sum of all resources across all of the monitored VMware hosts:

VMware Asset Summary   EDIT HELP

Number of Virtual Centers	0
Number of Clusters	0
Resource Pools	3
ESX Hosts	0 clustered, 1 non-clustered
Number of VMs	10 running, 13 total
Total Number of Physical CPU Cores	2
Total RAM	7.9 GB
Last Poll	38839 minutes ago
Platform	VMware

The remaining modules in **Virtualization Summary** are as follows:

- **VMware vCenters with Problems**
- **VMware Clusters with Problems**
- **VMware ESX Hosts with Problems**
- **VMware Guests with Problems**

The keyword in each of these modules is **Problems**. Any type of error, warning, threshold breach, state change, or any other type of problem will be shown in one of these four resources and allow administrators to be proactive in resolving issues before they become widespread.

If you click on one of the virtual hosts in the resource, it will display the following **ESX Host Details** view:

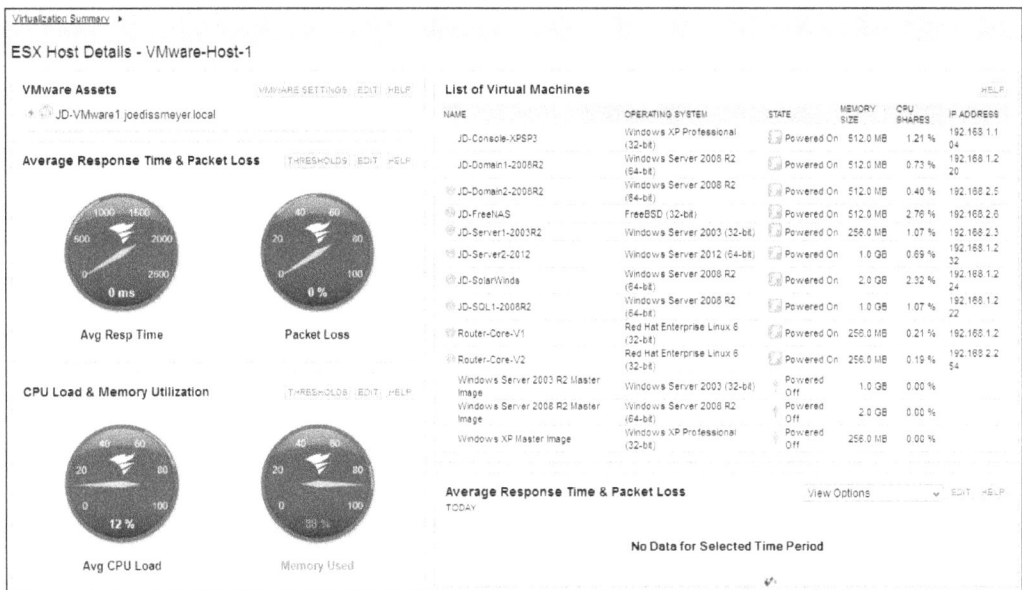

This view resembles the standard **Node Details View** page tremendously, but with several additions such as the following:

- **List of Virtual Machines**
- **ESX Host Details**
- **Virtual Machine CPU Consumption**
- **Virtual Machine Memory Consumption**
- **Virtual Machine Network Traffic**

The **List of Virtual Machines** resource is fairly straightforward. Displayed are all of the virtual machines that are attached, or running, on this virtual host as well as several details about the VM itself including the operating system type, memory assigned to the VM, CPU usage that the VM is using from the host, and the VM's IP addresses.

ESX Host Details displays all of the various details about the VMware host itself. Listed are its status, product version, and other tidbits of information:

The **Virtual Machine CPU Consumption** and **Virtual Machine Memory Consumption** views are a historical display for CPU and RAM consumption respectively for all virtual machines on this specific host over a period of time. From the following screenshot, you can see how the data is laid out. This information comes in handy when troubleshooting server issues, or when planning capacity.

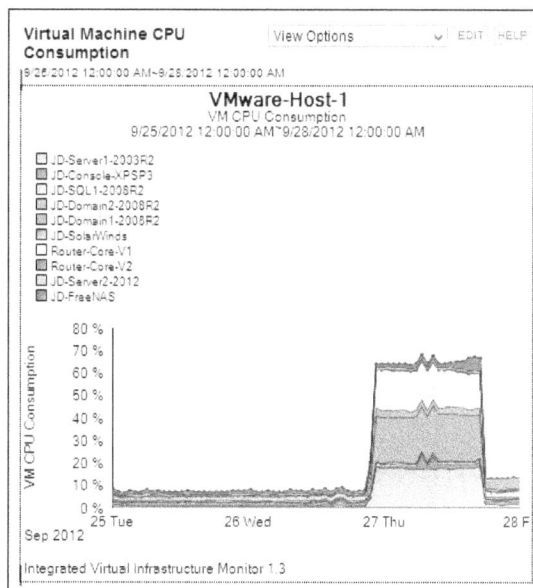

All other resources in this view are fairly self-explanatory and are easy to understand. As always, if you have a question on what data the module is displaying, click on the **HELP** button on the top-right corner of the module for a detailed explanation.

VSANs

Virtual Storage Area Networks (VSANs) are displayed in the **VSAN Summary** view under the **Network** tab.

There is a wide array of network storage options available including **Storage Area Network (SAN)**, **Virtual Storage Area Network (VSAN)**, **Internet Small Computer System Interface (iSCSI)**, **Network Attached Storage (NAS)**, and so on. All of these storage options connect to an isolated storage network, or converged network, using one of the two different types of network cabling: fiber or unshielded twisted pair copper. One of the most common ways that large storage arrays are connected to a virtualized infrastructure is by fiber channel interfaces, which uses fiber cabling. **Fiber Channel (FC)** is the de facto standard for connecting high-speed storage arrays to host servers.

A collection of fiber interface ports from a set of Fiber Channel switches form a virtual fabric called a virtual storage area network, or VSAN. Orion NPM specifically recognizes FC and VSAN devices and displays the monitoring data in the **VSAN Summary** view. A sample of the VSAN view is as follows:

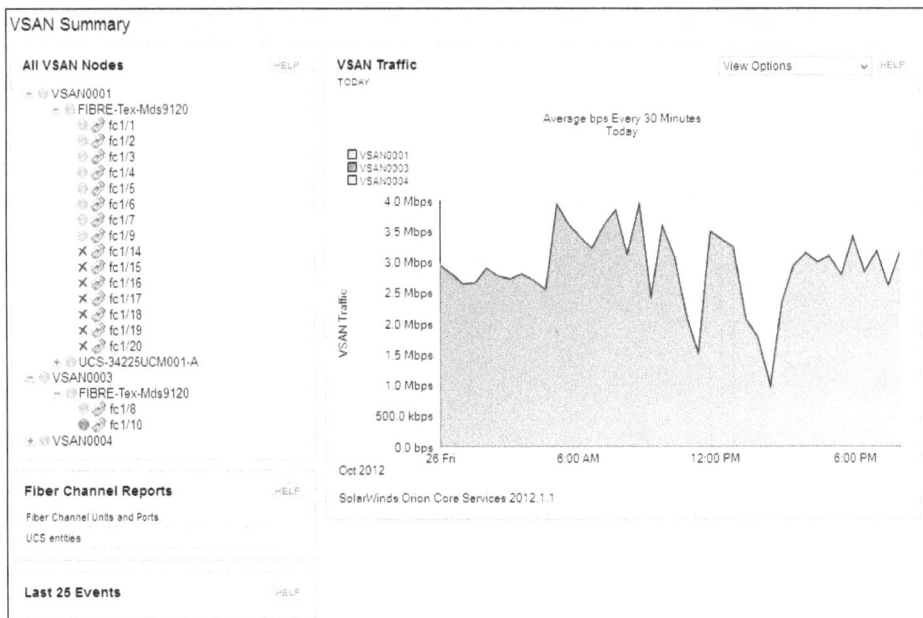

Orion NPM only applies Fiber Channel data storage to the VSAN view. All other storage types count as volumes attached to a monitored node.

The **VSAN Summary** view displays the following resources:

- **All VSAN Nodes**
- **VSAN Traffic**
- **Fiber Channel Reports**
- **Last 25 Events**

All VSAN Nodes may be the most commonly used resource in this view simply because of the fact that it displays all monitored VSANs, fiber switches, fiber channel **Host Bus Adapters (HBA)**, and each HBA interface. Of course, this resource also displays the state of each node and interface. The **All VSAN Nodes** resource displays all data in a hierarchical view first starting with the VSAN node itself, followed by the fiber switches or fiber channel host bus adapters that are a part of that VSAN node, and finally each fiber interface:

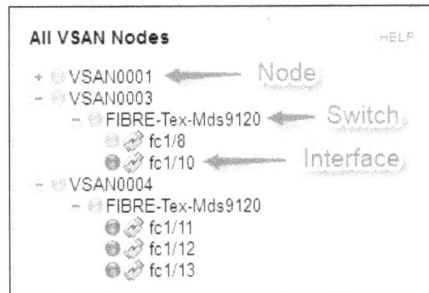

VSAN Traffic displays a chart of inbound and outbound traffic for VSAN over a period of time, the **Fiber Channel Reports** lists any custom reports for this view, and the **Last 25 Events** module displays any changes or problems that have to do with your monitored VSANs.

Clicking on one of the VSAN nodes will open the **VSAN Details** view:

All members of the VSAN, be it a FC switch, HBA, or interface, will be displayed in this view. Details about the VSAN are displayed in the **VSAN Details** module, with traffic and data details displayed in the **Total Bytes Transferred Every 30 Minutes** and **In/Out Errors and Discards** resources. Each of the last two resources display trends on the VSAN usage and hence can have custom reports created against this data, or alerts can be created if certain conditions are met. You can further drill down by clicking on one of the fiber channel ports, which will open the **Interface Details** page for that port.

> Notice that the storage capacity of the VSAN is not displayed anywhere in the **VSAN Summary** or **VSAN Details** view. This is because VSANs and fiber channels are a network transmission medium, not a storage medium! If you need to monitor the storage capacity, you will need to monitor it from the server that has the storage attached to it.

Cisco UCS

Cisco Unified Computing System (UCS) is Cisco's complete blade-style data center server platform that includes actual host server blades, embedded switching fabric, and virtualization and management software. Cisco UCS is specifically designed to run only VMware's ESXi (not ESX) hypervisor software. A single Cisco UCS chassis includes both the blade server hardware and the fabric switch. You can have multiple chassis when running Cisco UCS each with their own resources.

When monitoring your Cisco UCS infrastructure, you can view your UCS nodes from the **Home Summary** page or from the UCS view within the **Network** tab.

Each Cisco UCS chassis is a single node in Orion NPM, and each blade server (running VMware ESXi) is also one node. So, one chassis with four blades is a total of five nodes. Each component of the UCS chassis such as the power supplies, fans, blade servers, and fabric switches will be displayed as resources within the **Node Details View** page of the Cisco UCS system.

There are two pieces to monitor Cisco UCS. The first is the fabric switch within the UCS node, and the second is the actual VMware host running in a blade within the chassis. When monitoring Cisco UCS with Orion NPM, the blade servers running VMware ESXi are displayed as nodes in the **Virtualization Summary** page while the fabric switches and fiber channel interfaces are displayed in the **VSAN Summary** page.

Viewing the **Node Details** page of a Cisco UCS system should look very familiar to you by now since it has all of the same modules in this view. However, there are a few modules specific to Cisco UCS that you should be aware of. They are as follows:

- **List of VSANs**
- **Connectivity Unit Status**
- **UCS Overview**

Depending on the model of Cisco UCS hardware you are monitoring, there may be multiple VSANs configured. The **List of VSANs** view displays all of the VSANs available from within the chassis.

Connectivity Unit Status displays the status of VSAN nodes within the Cisco UCS chassis. Included is the node **NAME**, the VSAN's **World Wide Name (WWN)**, the **MODEL NAME** and **SERIAL** of the device, and the connectivity type (**CU TYPE**) as reported by the VSAN node.

Connectivity Unit Status					HELP
NAME	WWN	MODEL NAME	SERIAL #	CU TYPE	REVIS ID
MIA34225UCM001-A	20:00:00:05:9B:73:7A:C0	N10-S6100	SSI13510KJL	Switch	

UCS Overview displays all of the fabric interconnects and blade servers within the Cisco UCS chassis. Each blade server that is monitored by Orion NPM is listed under the parent chassis. Certain models of Cisco UCS have multiple chassis and will be displayed in the **UCS Overview** model provided they are being monitored by Orion NPM.

UCS Overview HELP

Fabric Interconnects

NAME	RESPONSE TIME	PERCENT LOSS	STATUS
switch-A	N/A	N/A	N/A

Chassis & Blades

NAME	RESPONSE TIME	PERCENT LOSS	STATUS
chassis-1			
blade-1	N/A	N/A	N/A
blade-2	N/A	N/A	N/A

Universal Device pollers

Out of the box, Orion NPM allows you to monitor just about every SNMP-enabled device on your network. However, there may be cases where the default views and monitors that Orion NPM displays for your monitored device are not sufficient. For example, you may be more interested in the battery life remaining in a UPS battery backup instead of its interface throughput. Another example would be you want to view a list of virtual interfaces on a F5 BIG-IP node. This is where Universal Device Pollers come into play.

Universal Device Pollers allow you to create custom monitors for almost any statistic provided by SNMP based on the device's MIB and OID that is outside of the standard monitoring parameters. **Management Information Base (MIB)** is a hierarchical database of **Object Identifiers (OIDs)**. SolarWinds Orion NPM has its own MIB database that is regularly updated and contains OIDs for thousands of makes and models of network devices.

> Universal Device Pollers only apply against nodes that can be monitored using SNMP.

Let's start by looking at the Universal Device Poller application. It is launched from the Windows Server that Orion NPM is installed on and is available in the Start menu.

Once the application appears on the screen, we are going to take a look at how to create new pollers and how to apply them to devices that we are monitoring with Orion NPM. The Universal Device Poller application will look like the following screenshot:

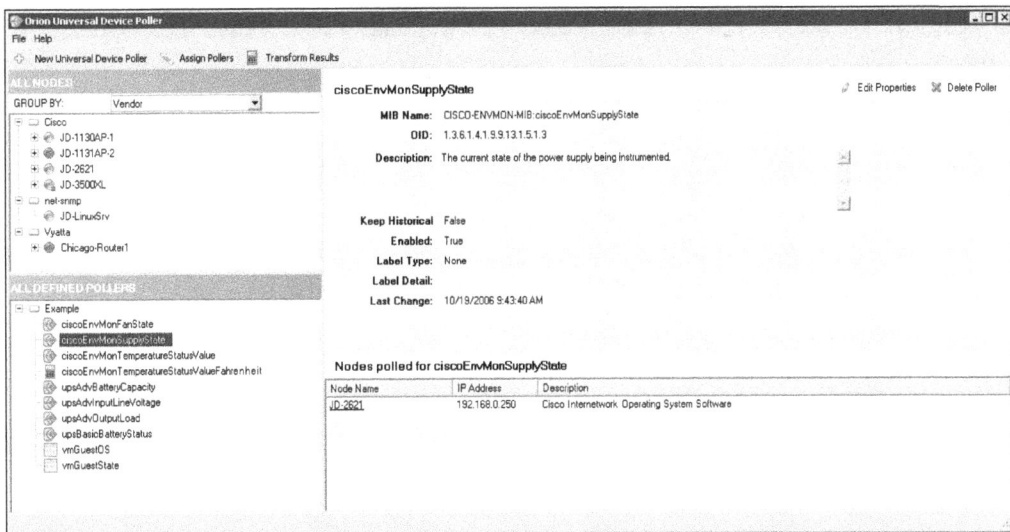

On the left-hand side of the application window is the **ALL NODES** pane. This is a list of all of the nodes that you are currently monitoring with Orion NPM.

Underneath is the **ALL DEFINED POLLERS** pane, which lists all of the available Universal Device Pollers. Orion NPM provides several examples of pollers out of the box which are in the **Example** folder. By default, none of these example pollers are assigned to a node.

When clicking on a poller in the **ALL DEFINED POLLERS** pane, the main display pane on the right-hand side displays the properties of the poller, as well as which nodes are being monitored by it. You can also delete and edit the properties of the poller via the links on the upper-right hand section of the display pane.

With the Universal Device Poller, you can create the following three different types of pollers:

- Individual value poller
- Transforms
- Table-based pollers

Creating an individual poller

To create a new individual value poller, perform the following steps:

1. Click on the **New Universal Device Poller** button.
2. Define the **OID** value in the textbox:

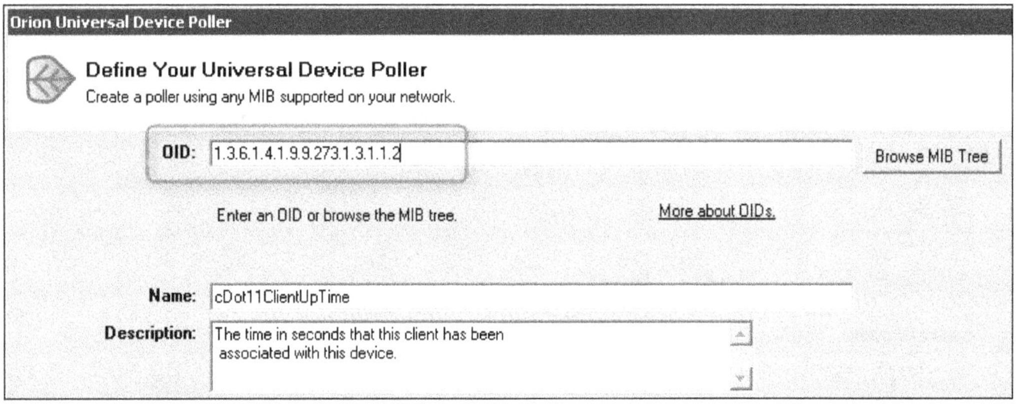

In this example, I already knew the poller that I wanted to use. As it was typed in, the **Name** and **Description** of the OID was auto-populated.

If you do not know the OID value, click on **Browse MIB Tree** to manually locate the OID value you wish to assign. Once you find the OID, select a node that you want to apply the poller against and click on **Test**. In the example, I am testing the **dot11ClientUpTime** OID against one of my access points to tell me the up time of all of the connected clients:

If the test was successful, click on the **Select** button to continue. Otherwise, locate another OID to test against. As you can see, my test was successful, so I will click on the **Select** button to continue:

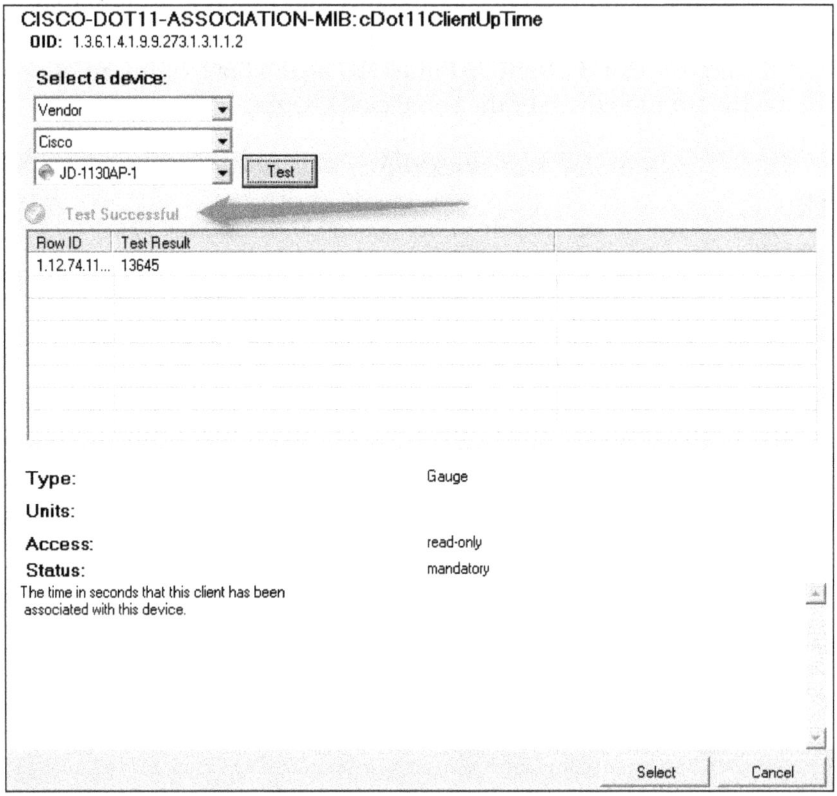

3. Choose if you want to keep the poller's historical data in the Orion database, if you want to enable the new poller or disable it, and choose a group name for the new poller. Click on **Next**.

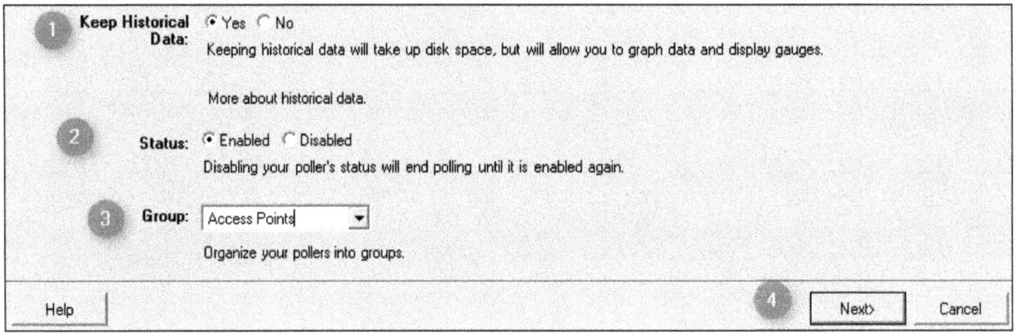

In the example, I am going to keep all historical data related to this new poller and I will also enable it immediately. Also, I defined a new name for the poller group titled **Access Points**.

4. Choose which node, or nodes, you wish to apply the new poller to. Click on **Next** to continue.

5. Next, choose a label for the poller. The label is what will be displayed for the poller data in the Orion dashboard. You can use one of the interface names from Orion by selecting a label from one of the OID's table columns, or defining your own label by choosing **Use a custom label**. In this case, I am going to choose the label from the OID's table column **cDot11ClientUpTime**. Click on **Next** when you are ready to continue.

6. The last step in creating a new universal device poller is to choose whether or not to include the data in the Orion dashboard. You can choose **Yes** or **No**. When choosing **Yes**, you need to decide which web view the data will display on. Click on **Finish** to create the poller.

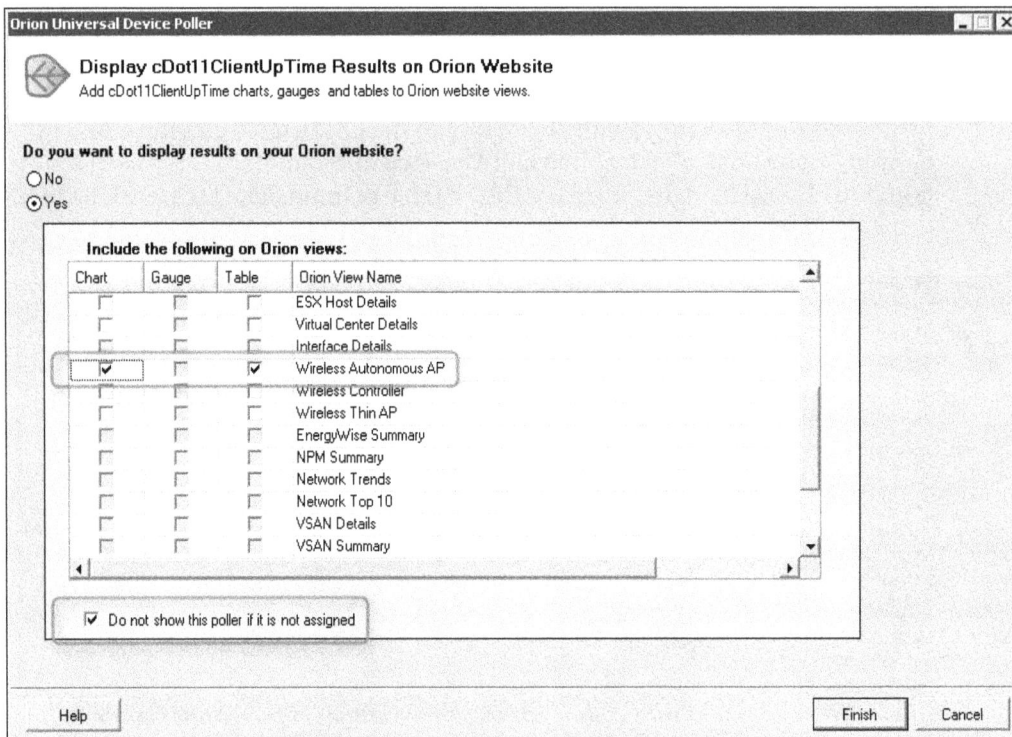

Since I am creating the new poller for a Cisco access point, I am going to apply this new poller to the **Wireless Autonomous AP** view. Also, I selected the checkbox for **Do not show this poller if it is not assigned**. This option will hide the resource view for this poller in the Orion dashboard if the poller has not been applied to a node. This means that when I open the **Wireless Autonomous AP** view of the access point, the poller resource will be displayed. The resource will not display for any other nodes in the same web view since the universal device poller has not been applied to them.

7. The new Universal Device Poller has now been created and can be seen on the **ALL DEFINED POLLERS** view pane in the group you defined in the wizard. You will also see which nodes the new poller has been applied to in the main display pane.

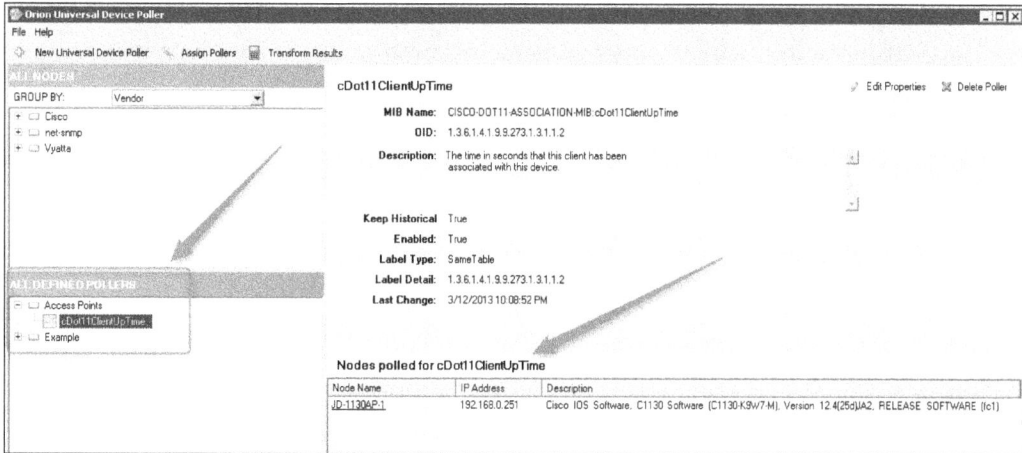

In this example, I applied the new access point poller against the **JD-1130AP-1 node**. When I go back to the Orion dashboard and click on the access point, I can see the new poller. But I need to convert this value into something that makes sense. To do this, we use a transform poller.

Creating a transform poller

There are some situations where you will need to convert data from a universal device poller into a different format. In the example from the previous section, the **14968** value that the new poller is displaying does not appear to present anything of value. To convert this data into something meaningful, you create a transform poller.

To create a new transform poller, perform the following steps:

1. Click on the **Transform Results** button at the top of the window.

2. The **Transform Results Wizard** appears. The first step in creating a transform poller is similar to creating a new Universal Device Poller. Define the poller's name, enter a description, choose to keep or dump historical data in the Orion database, choose to enable or disable the poller, and select a group to save the poller to. Alternatively, you can type in a new group name. Click on **Next** when you are ready to continue.

> You can only use alphanumeric characters in a poller's title. You cannot use any type of symbol or special character, including dashes (-) and underscores (_).

3. Next, click on the **Add Poller** button and choose the poller that you want to transform. In the following example, I selected **cDot11ClientUpTime**:

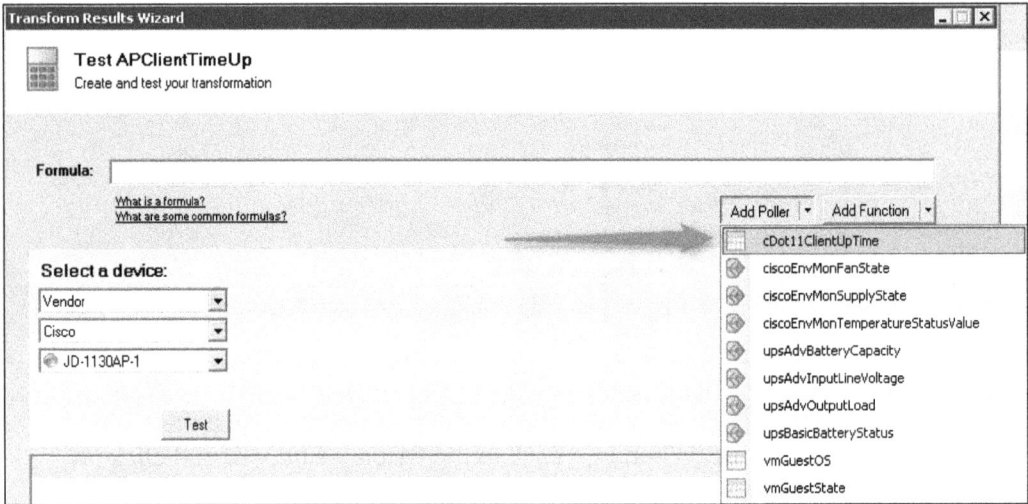

4. The formula will populate in the textbox. Now, add in the proper functions to calculate the result you wish to display and then click on the **Test** button to verify the results. Click on **Next** to continue.

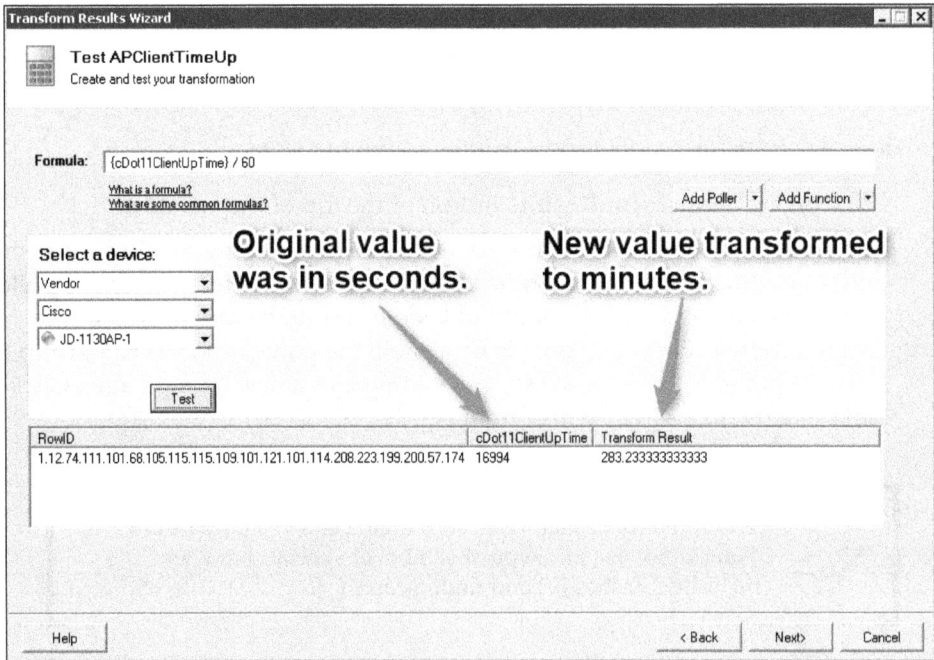

To convert my new poller **cDot11ClientUpTime** from seconds to minutes, I need to divide the poller result (indicated by brackets) by 60. After testing the results, the value **16994** seconds was correctly transformed to **283.23** minutes.

5. Next, choose which nodes to apply the transform poller to. Click on the **Test** button to verify that the poller works as expected and then click on **Next**.

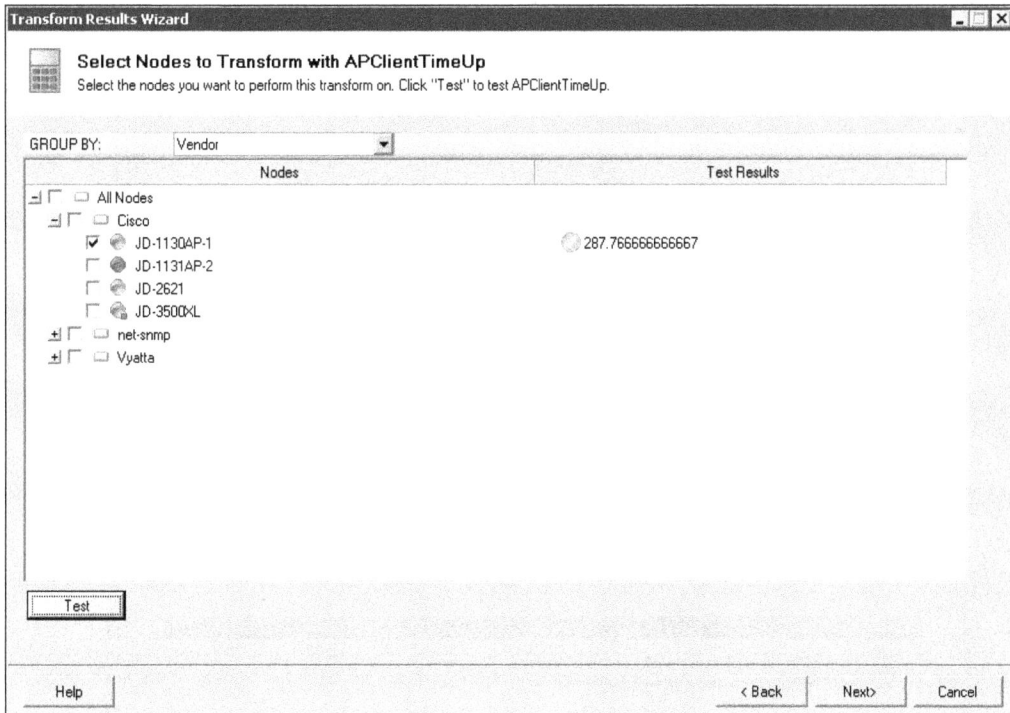

6. Choose a label for the poller. In this example, I will choose a custom label named **Client Up Time**. Click on **Next**.

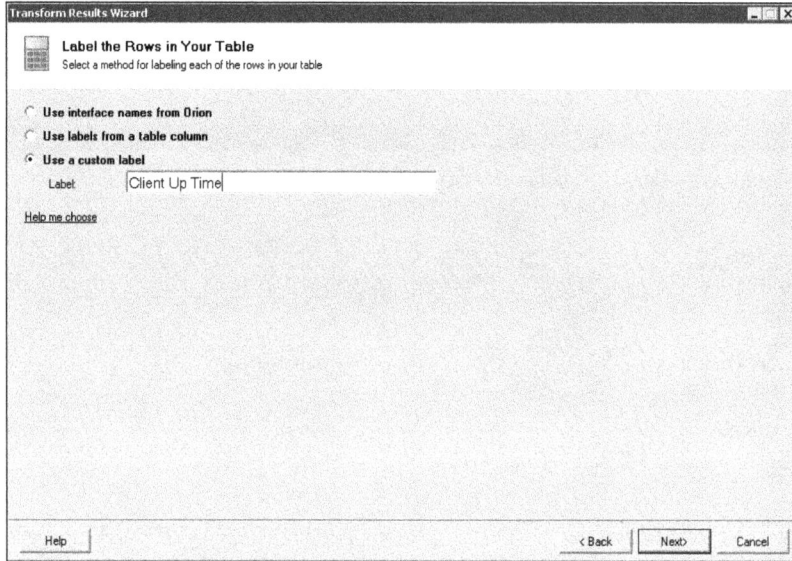

7. The final step is to choose if you want to include this poller as a resource in the Orion dashboard. Choose which view, or views, you want to apply the poller to and then click on **Finish**:

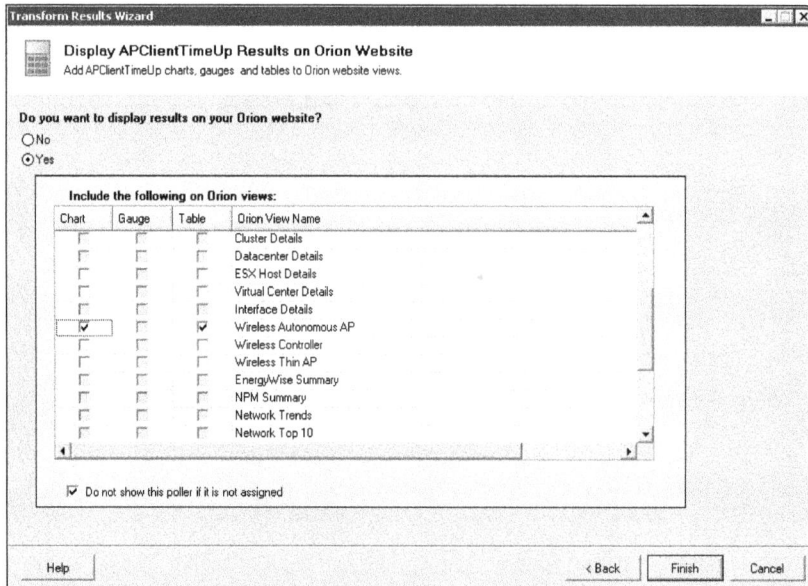

In this example, I chose to display the transform poller in the same web view as the original poller. So now, when I go back to the Orion dashboard and refresh the **Wireless Autonomous AP** view for the access point, the new transform poller will be displayed:

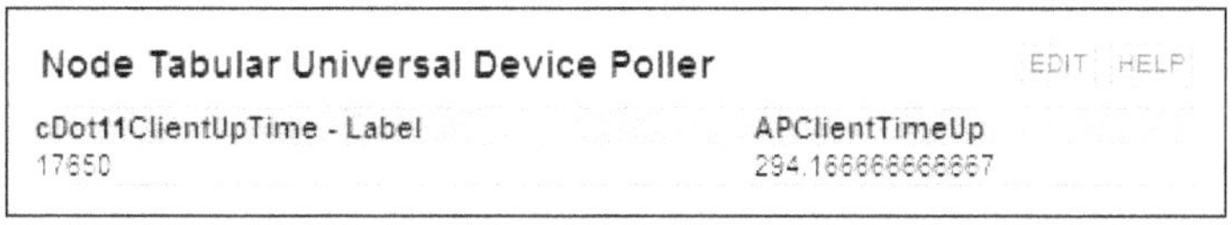

```
Node Tabular Universal Device Poller              EDIT  HELP

cDot11ClientUpTime - Label          APClientTimeUp
17650                               294.166666666667
```

However, the resource title makes no sense and I want to include the new transform poller data. In order to change the resource module contents, perform the following steps:

1. Click on the **EDIT** button.

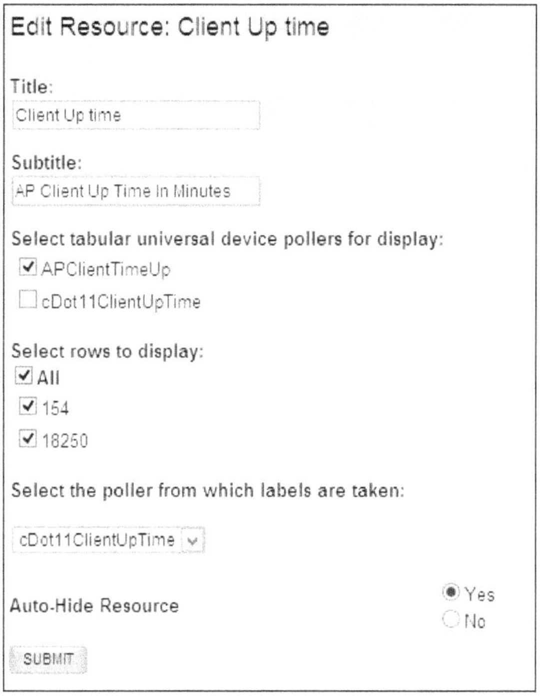

```
Edit Resource: Client Up time

Title:
Client Up time

Subtitle:
AP Client Up Time In Minutes

Select tabular universal device pollers for display:
  ☑ APClientTimeUp
  ☐ cDot11ClientUpTime

Select rows to display:
  ☑ All
    ☑ 154
    ☑ 18250

Select the poller from which labels are taken:

  cDot11ClientUpTime  ▾

                                      ⦿ Yes
Auto-Hide Resource                    ◯ No

  SUBMIT
```

2. Remove the checkbox next to the original Universal Device Poller **cDot11ClientUpTime** so that it does not display in the view. Also, change the title and subtitle. After clicking on **SUBMIT**, the resource displays easy-to-read data.

Client Up time	EDIT HELP
AP CLIENT UP TIME IN MINUTES	
cDot11ClientUpTime - Label	APClientTimeUp
154	2.56666666666667
18250	304.166666666667

This is only one example of how you can create your own Universal Device Pollers. The last feature I will show you regarding Universal Device Pollers is the import and export feature.

Exporting and importing pollers

The Universal Device Poller allows you to export pollers to a file. It also allows you to import pollers that may have been shared with you. There are two reasons why you would want to import and export universal device pollers:

- For backups
- For sharing

Universal Device Pollers are not stored in the Orion database, so it is a good idea to regularly export your pollers. Another reason for exporting pollers is to share them with another team that may be handling their own Orion NPM system. A third reason is to share them with a colleague, or to share them on SolarWinds' online community, Thwack, at `http://thwack.solarwinds.com/welcome`.

> We discuss the Thwack community and Orion backups in *Chapter 8, Maintenance*.

Exporting pollers

To export your pollers, perform the following steps:

1. Click on **File**, and then choose **Export Universal Device Pollers**.

2. Select one or multiple pollers, and then click on the **Export** button:

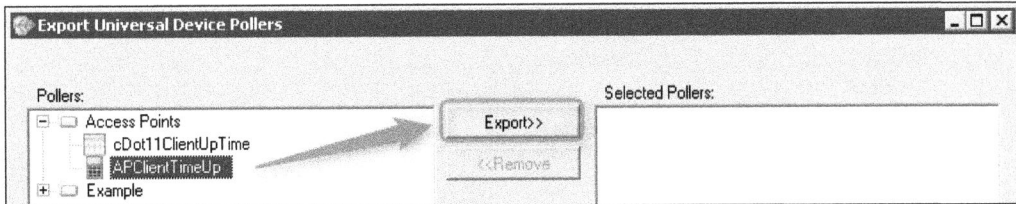

3. When you have selected all of the pollers you wish to export, click on the **Save** button.

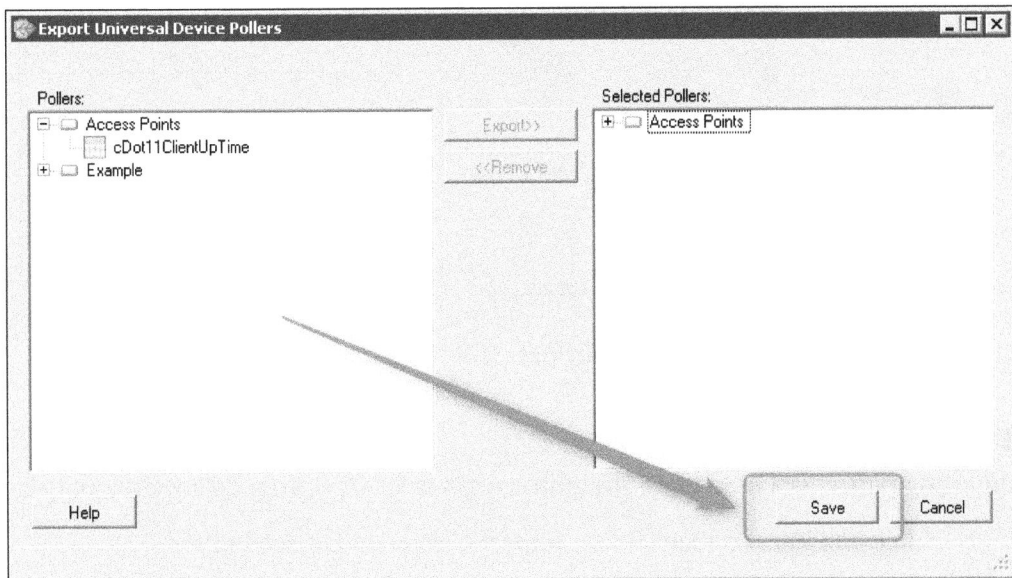

4. Choose a name and location for the exported poller, and then click on **Save**:

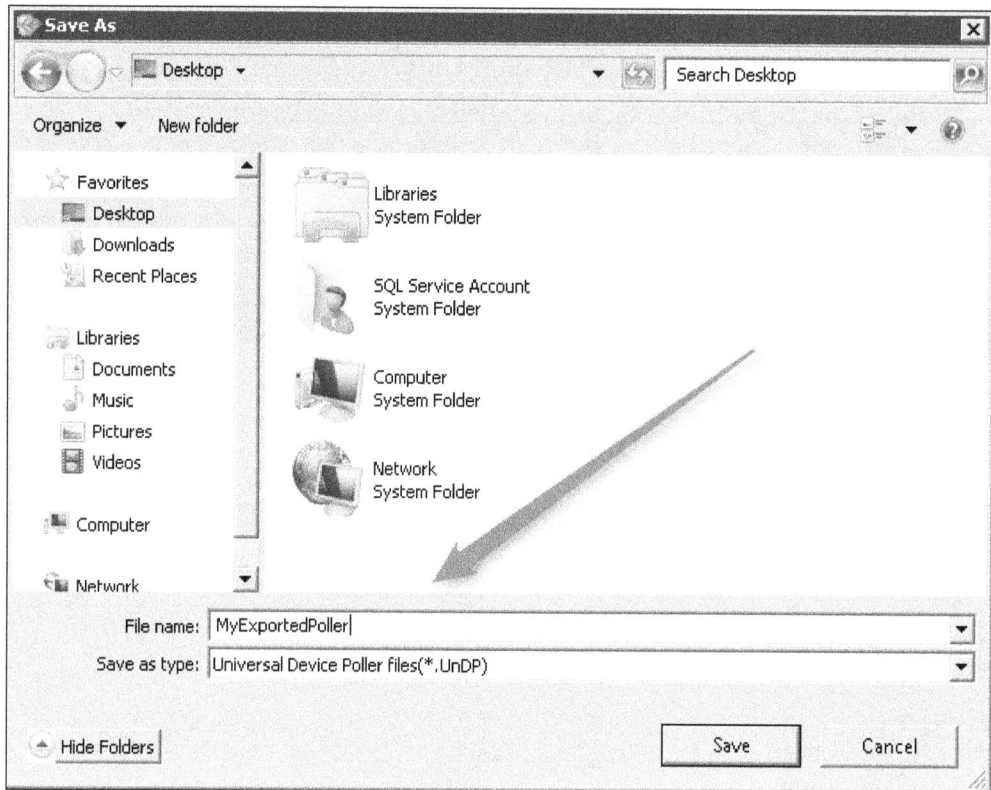

Orion NPM saves the file in UNDP format. This file can now be copied and shared.

Importing pollers

Importing pollers is a little more straightforward than exporting. To import a poller:

1. Click on **File**, and then choose **Import Universal Device Pollers**.
2. On the **Import Universal Device Pollers** window, click on the **Open** button and locate the UNDP file.

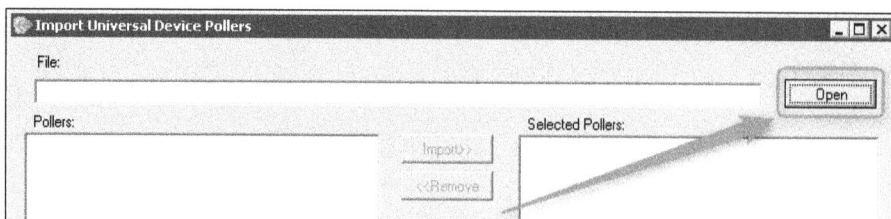

3. Highlight all of the pollers you wish to import and click on the **Import** button. Or, you can import all of the pollers at once by highlighting the folder. Click on **OK** to finish:

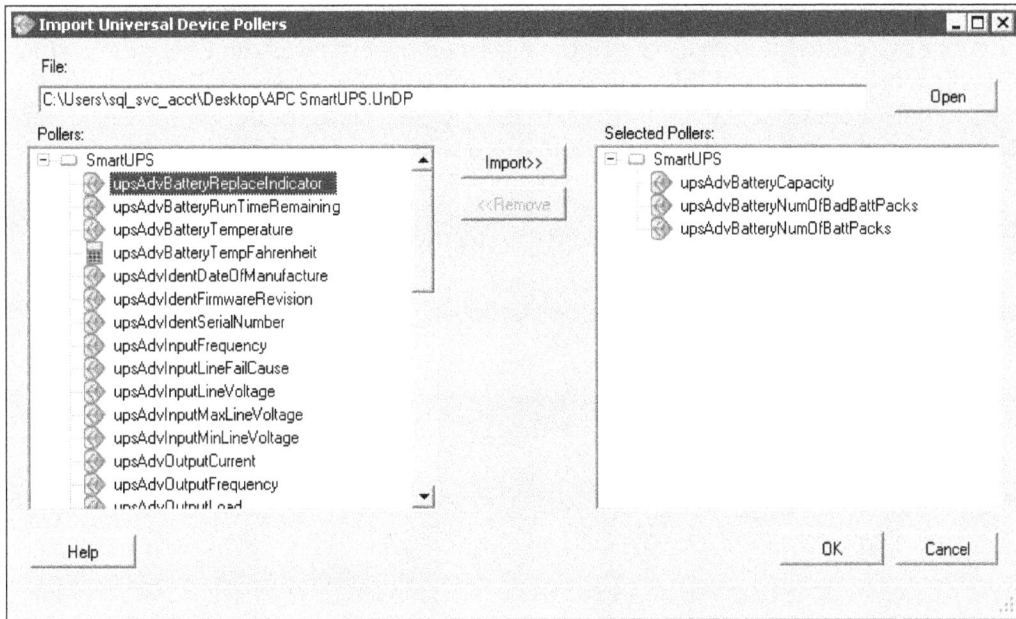

In this example, I imported a Universal Device Poller for APC SmartUPS uninterruptable power supplies. The imported pollers will be listed in the **ALL DEFINED POLLERS** view pane.

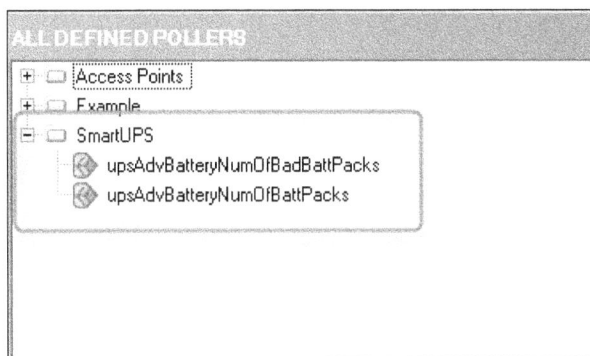

Summary

This marks the end of the chapter! Are you having fun yet? Good! In this chapter, we discussed the basics of server monitoring, SNMP and WMI configuration for servers, virtualization monitoring, VSANs and Cisco UCS, and Universal Device Pollers. I'm sure that by reading this chapter you have learned quite a bit about how monitoring a server works in Orion NPM, as well as how the different server and virtualization technologies operate. To me, this is one of the biggest strengths of Orion NPM in that it gives you in-depth details about a network device and displays the data in a single pane of glass that makes sense.

In the next chapter, we are going to fully discuss alerts, which is one of the largest features of Orion NPM.

6
Setting Up and Creating Alerts

Your standard modern automobile is a very sophisticated piece of machinery. There are many different bells and whistles that come standard with a new car, which may include a DVD player, built-in GPS, and OnStar service. Some features that now come with a new car would make *KITT* jealous (gratuitous *Knight Rider* reference). All vehicles also have a complicated engine with hundreds of moving parts that need to work together in order for it to work properly. The engine needs to be properly calibrated, oiled, and fuelled in order to operate. In addition to the core parts of the vehicle, it has a computer and gauge system which monitors many aspects of the entire system. This computer and gauge system allows you to be proactive in the maintenance and operation of the vehicle. Every time this system warns you about a possible problem with the vehicle, it is called an **alert**.

Now imagine that you no longer have an alert system for your vehicle or any gauges in the heads-up display. How can you be proactive in making sure that you are following the law regarding speed limit? How can you be sure that you have enough fuel to drive to work? Without an alert system available, it would be very difficult to know what needs to be maintained in your vehicle without excessive manual check-ups by you or a skilled automobile technician.

Take this same concept and apply it to your computer network. Without an alert system, how would you know if there was a problem? How would you know if a node was down without one of your customers calling you first? Now imagine if there were no alerts about the core Internet infrastructure. There would probably be little or no Netflix, or Facebook, or Xbox Live! Oh the horror! The fact of the matter is that, without a network monitoring and alert system, you wouldn't know if there was a problem, period.

This chapter is strictly focused on discussing the alert system built into Orion NPM and covers alerts generated by Orion NPM. In it, we will discuss how to set up basic alerts, advanced alerts, and manage various alert configurations.

Orion NPM alerts

Alert functions helps to make Orion NPM a much more useful system. Orion NPM has two different types of alerts that can be configured and activated, basic alerts and advanced alerts.

A **basic alert** is one that e-mails an administrator if a node goes down, an interface has a high packet loss, or if a node has rebooted. An **advanced alert** is one that is generated when Orion NPM detects more than a defined amount of clients on an access point, or if a threshold has been exceeded for an extended amount of time, and other non-standard types. Many different types of alerts can be created and every one of them can have a variety of triggers. It all depends on what information you need to be alerted on in the event of a change.

Alert acknowledgement

Every alert is recorded in the Orion NPM database as an event and is displayed in the **Alerts** view on the Orion dashboard. Every alert will be listed in this view until it has been acknowledged by an administrator.

By default, the last 250 alerts are displayed in the list on the **Alerts** view. A search filter is embedded at the top of the view which allows administrators to search against dates, times, and monitored resources to find alerts within a specific time frame. From the preceding screenshot, you can see that there are several alerts waiting to be acknowledged. Displayed is the time of the alert, the alert title, the type of the alert (advanced or basic), and what resource the alert was triggered against. The very top alert informs us that the node **FreeNAS** is currently down, which is correct because I personally shut it down before writing this chapter! To acknowledge an alert, perform the following steps:

1. Place a check mark next to the alert on the left-hand side of the **Alerts** view then click on the **Acknowledge Alerts** button.

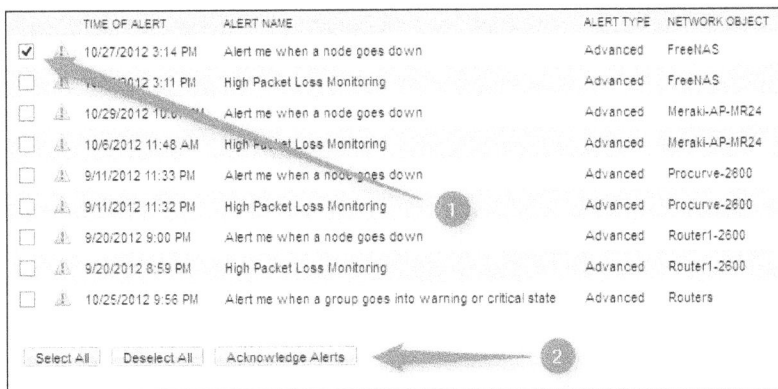

2. Write an administrative note in the textbox and click on **OK** to save.

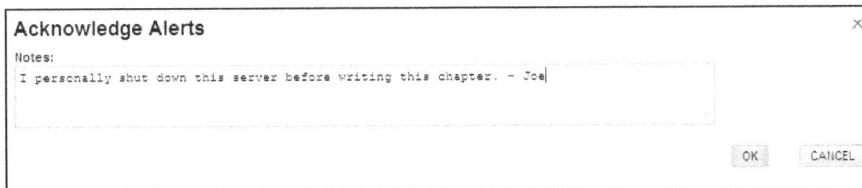

3. The alert is now acknowledged and has been removed from the list.

You can choose more than one alert to acknowledge at the same time by placing a check mark next to each one, or you can click on the **Select All** button.

An alert will not simply "go away" when the problem has been resolved or if the node/resources is operating as normal. An administrator must acknowledge the alert in order for it to disappear from the **Alerts** view, or the alert must be suppressed when it is back to normal. Alerts are designed to work this way because Orion NPM allows you to build reports based on alerts.

Preconfigured alerts

When Orion NPM is first installed, several alerts are enabled which allows administrators to monitor and alert against their network right off the bat. The following alerts are pre-configured, but not enabled, out of the box:

- When a node, interface, wireless access point, or group goes down
- When a node reboots
- When a device experiences high packet loss, high response time, or high transmit percent utilization
- When a group goes into warning or critical state
- When a managed node has not been polled during the last five tries
- When a managed node's last poll time is 10 minutes old
- When a polling engine has not updated the database in 10 minutes
- When a rogue access point is detected
- When a wireless access point has more than 10 clients
- When someone shuts down an interface
- On high bandwidth utilization by an access point
- On a Cisco iOS version or image family change

> Even though these alerts are preconfigured when Orion NPM is installed, it is still up to you to select which ones you want enabled.

Configuring alerts is done from the SolarWinds Orion NPM server itself. First, let's discuss enabling basic alerts.

Configuring basic alerts

Configuring basic alerts is a great way to get started with understanding how the alert system works. We configure a basic alert by using the Basic Alert Manager application which is installed on the Windows Server that SolarWinds Orion NPM is installed on.

To access the Basic Alert Manager application perform the following steps:

1. Log into the SolarWinds Orion NPM server.

2. Navigate through **Start | All Programs | SolarWinds Orion | Alerting, Reporting, and Mapping | Basic Alert Manager**.

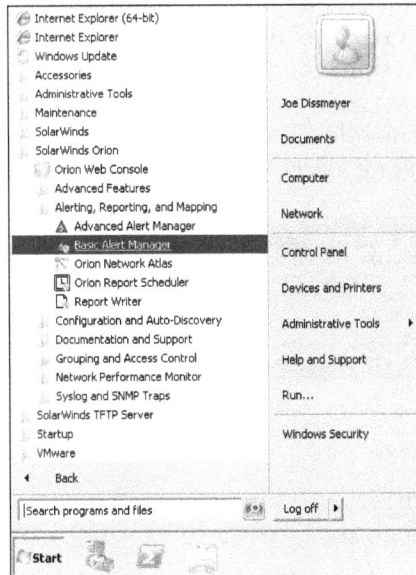

When the **Basic Alert Manager** application launches, the **Quick Start** screen is always displayed first. To get started, click on the **Configure Alerts** button in the main window.

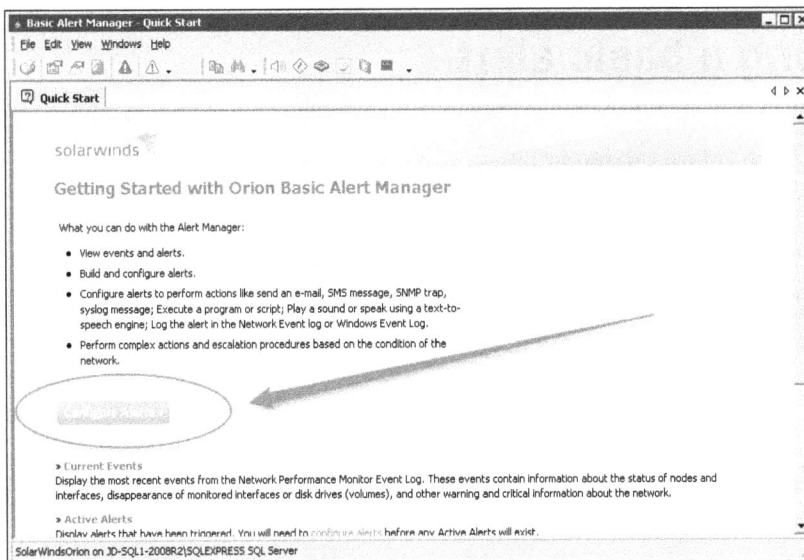

The **Configure Alerts** window appears and lists all of the basic alerts already set up in Orion NPM. All of the alerts are disabled by default because the alert actions must be edited one by one. They are added to the alert list as examples. To enable an alert, simply place a check mark next to its title and remove the check mark to disable it. To temporarily disable actions on all alerts in the list, place a check mark next to the **Temporarily Disable all Actions for All Alerts** option. You may want to disable all actions on the Orion NPM server in case you are performing regular maintenance, such as a reboot. From this window, you can create new alerts, delete or edit existing alerts, and test fire any existing alerts.

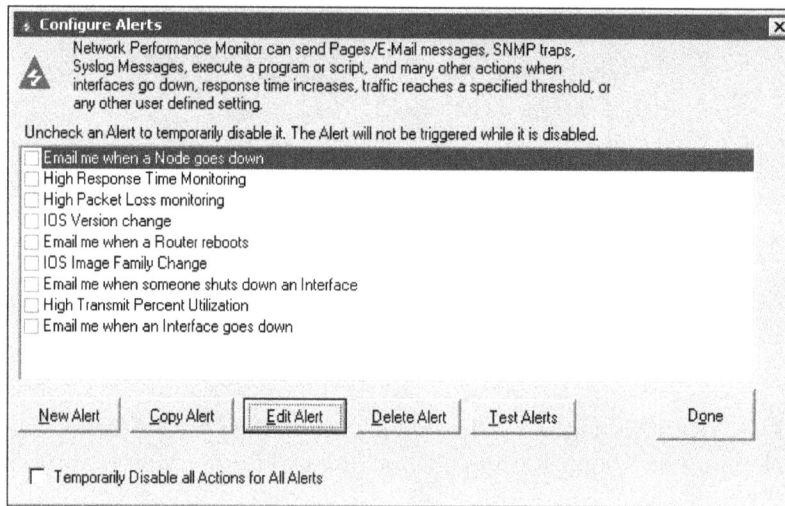

Building a basic alert

Building a basic alert involves the following process:

1. Give the alert a name or a title and enable or disable it.
2. Select which properties to monitor.
3. Choose which objects to apply the monitor to.
4. Define the alert trigger.
5. Define the date and time to enable or disable the monitor.
6. Define alert suppression.
7. Define the action to be taken when the trigger is hit.

Let's walk through the alert building process by setting up a new alert from scratch. To create a new alert, click on the **New Alert** button. The **Edit Alert** window will appear.

From the **General** tab, give the alert a title and choose to enable the alert or not.

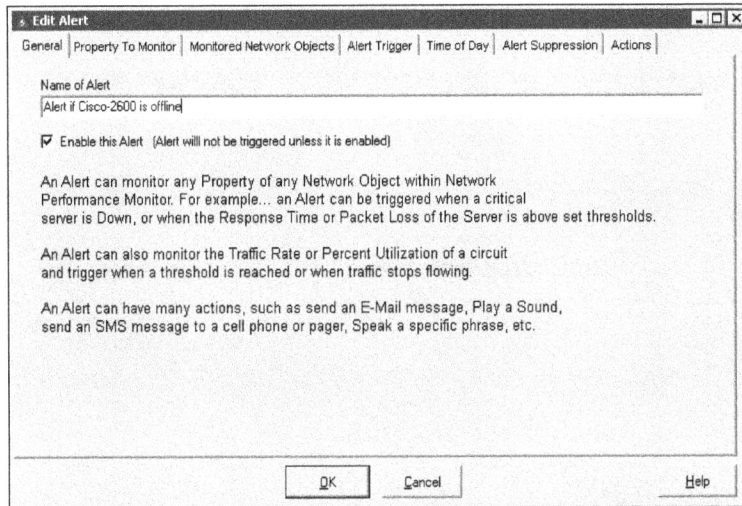

Next, click on the **Property to Monitor** tab and choose one or multiple monitored properties to alert against. There are a great deal of various options in this tab. In this example, I am choosing **Status** under the **Node Status** section, as I only want to alert if a node is down.

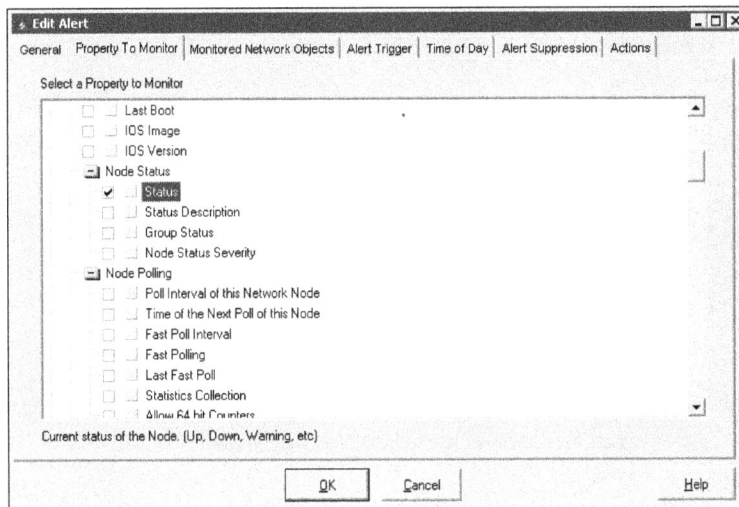

Click on the **Monitored Network Objects** tab and choose one or multiple nodes to apply the alert to. You can choose all objects if you wish, or choose a few. When ready to continue, click on the **Alert Trigger** tab.

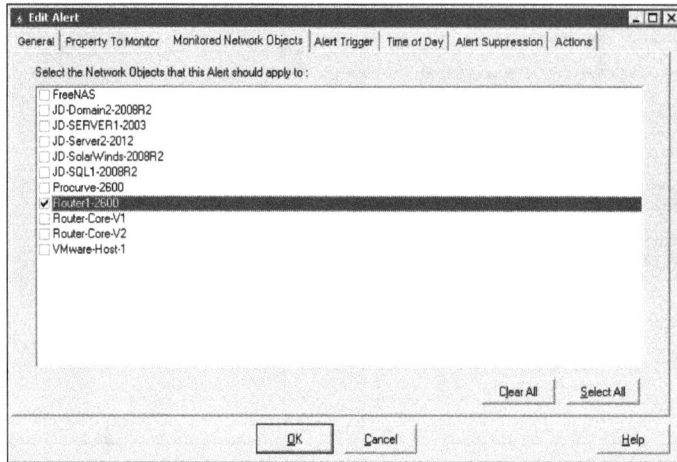

The **Alert Trigger** tab is where you define what the cause of the alert will be. Choose the status type for an alert to be triggered or reset. The easiest thing to do here is to place a check mark in the **Send Alert any time Status changes** option which will grey out the entire tab. Yes, it is possible that this option could generate excessive reports in a dynamic network environment, but for many, it is a good choice. For this example, I want to know when the node is down, and I want to reset the alert once it comes back up.

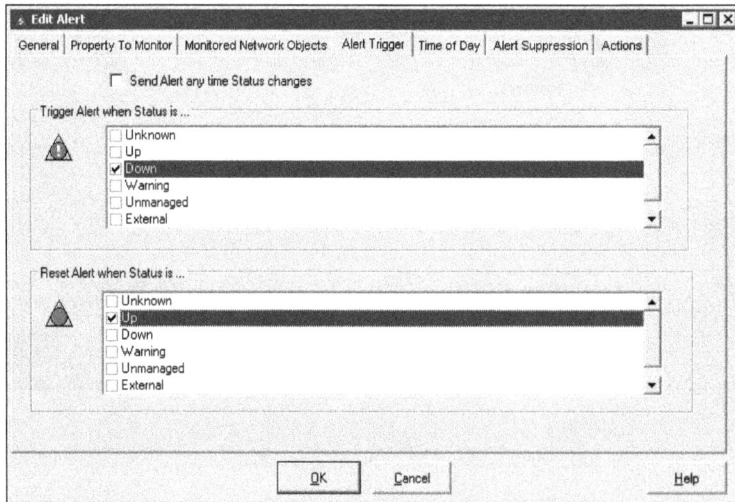

In the **Time of Day** tab, choose the time and day to apply the alert. The default configuration is 24 x 7, every day of the week. There may be certain days or times where you have maintenance hours and you don't want to generate any alerts during that time frame, so make your configuration changes accordingly.

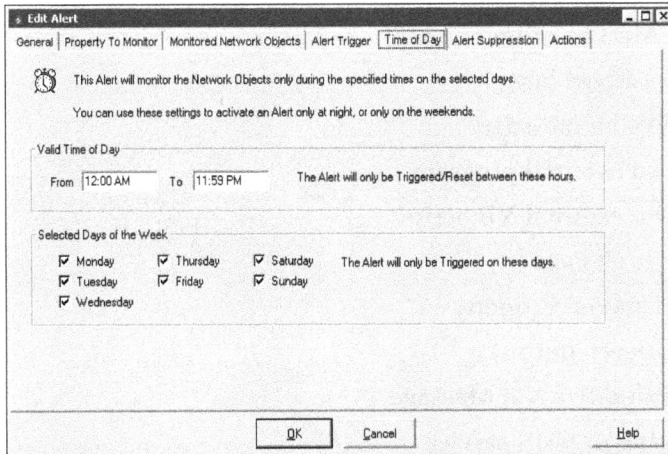

The **Alert Suppression** tab allows you to stop an alert action from occurring based on certain circumstances. For an example, you may have two servers that are part of a high-availability cluster pair running in active-passive mode. If the active node fails over to the passive node, you may want to suppress the Orion NPM alert as long as the second node is still online. Another example of why you may want to suppress an alert is you have two network load balancers where the bandwidth threshold has been exceed on one, but the second has a low bandwidth utilization. The options are virtually endless here. The default option is **Do not configure Alert Suppression for this Alert**.

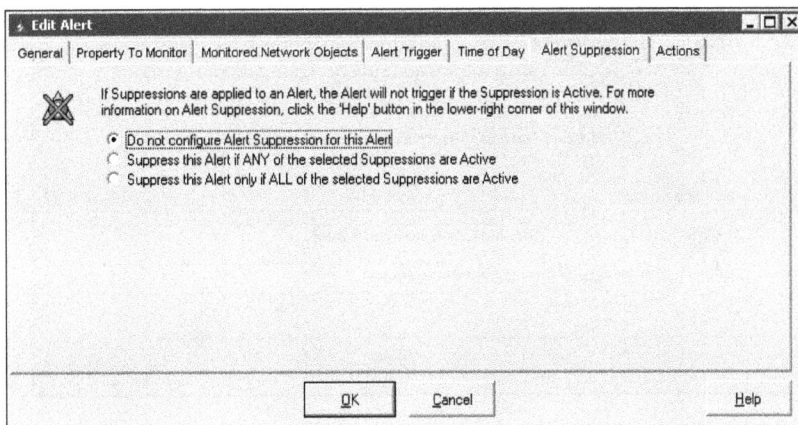

The final step in creating a basic alert is to select one or multiple actions to take when an alert trigger has occurred. Some of the actions that can be taken are as follows:

- **Send an E-Mail / Page**
- **Play a sound**
- **Log the Alert to a file**
- **Windows Event Log**
- **Send a Syslog message**
- **Execute an external program**
- **Execute an external VB Script**
- **E-mail a Web Page**
- **Change Custom Property**
- **Text to Speech output**
- **Send a Windows Net Message**
- **Dial Paging or SMS service**
- **Send an SNMP Trap**
- **Get or Post a URL to a Web Server**

The first option of sending an e-mail is the most common one. You can send an e-mail alert via SMTP to any e-mail address of your choice, as well as pager addresses. Choosing this option will open the **Edit E-Mail/Page Action** window.

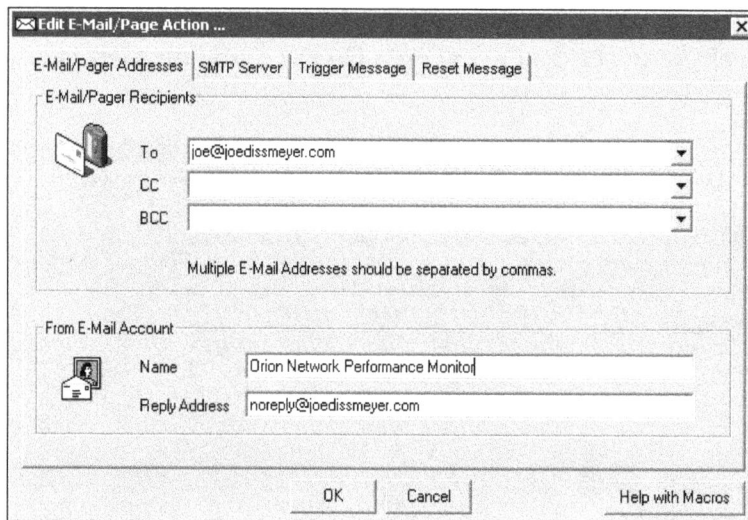

In the **E-Mail/Pager Addresses** tab, define the addresses that the alert will be
e-mailed to as well as a simple reply address. In my experience, it is always a good
idea to use a basic `noreply@emailaddress.com` address for public alerts. If you are
a managed service provider, a good alternative might be your company's inbound
support e-mail address (such as `support@buzzinga.com` or `help@walowizard.net`).
Click on the **SMTP Server** tab to continue.

Next, configure your SMTP server settings. All of the common SMTP configuration
options are available here and almost all public (free) and paid (Google Apps,
Exchange Server, Zimbra, and so on) e-mail solutions offer some type of SMTP service.

The **Trigger Message** tab allows you to change the contents of the subject line and body of the e-mail notification that will be sent. Orion NPM alerts support macros, or item-specific details from the Orion database that start with the dollar sign ($). In the preceding example, the **NODENAME** and **STATUS** macros are added.

> You can change these macros to anything within the Orion database. Click on the **Help with Macros** button for detailed information.

The final settings are on the **Reset Message** tab. This is the same as the **Trigger Message** tab, but this is the notification that will be sent out when the alert has been reset.

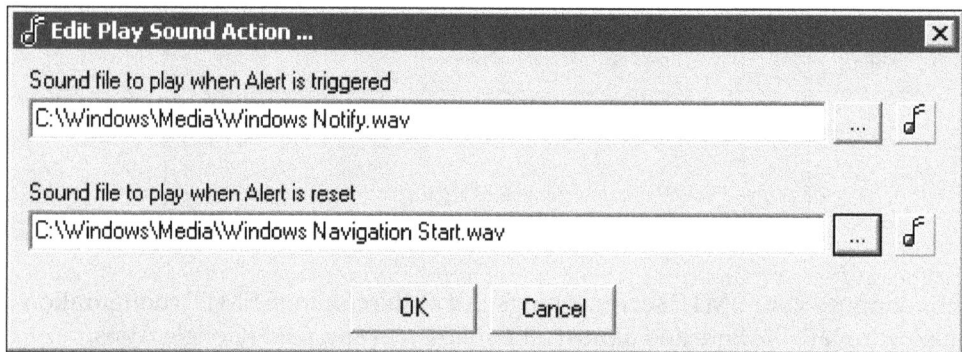

The **Play Sound Action** option will play a sound on the Windows Server when the alert is triggered and/or when the alert is reset. This can be a great option for "war rooms" or support desk areas where an audible alert is helpful. Orion NPM only supports WAV files so that Gangnam Style MP3 file won't work.

> Sound in Windows Server 2003 and Windows Server 2008 (R2) is disabled by default. In order to enable sound in Windows Server 2003, you must install the hardware audio drivers. To enable sound in Windows Server 2008 and 2012, you need to enable the **Desktop Experience** feature from **Server Manager**.

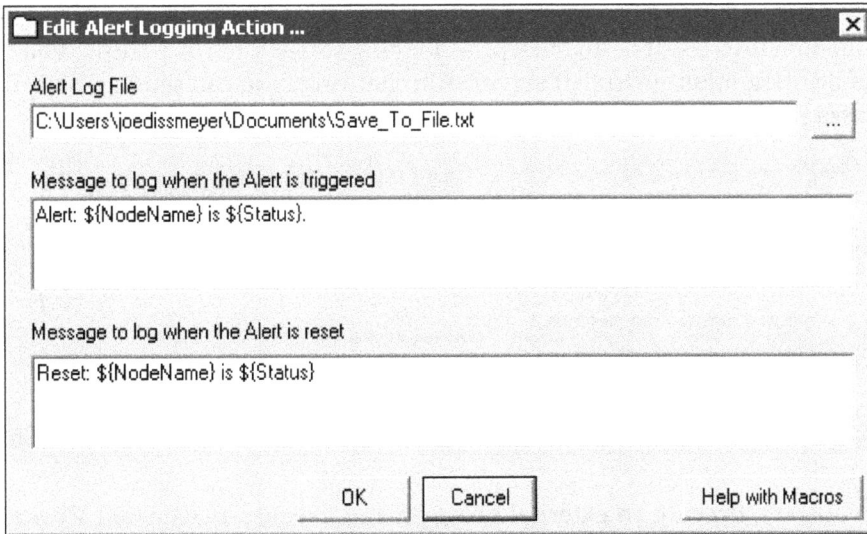

Logging the alert to a file will "dump" a text file containing the same type of text from an e-mail alert to a location of your choosing. If you export the file to a network folder, make sure the service account running SolarWinds Orion NPM has permissions to the network folder location.

Windows Event Log! Why the heck would you want to place an Orion alert in the Windows Event Log?! For reporting reasons of course, as well as company compliance reasons I imagine. Placing well, I'm sure there are reasons. First choose if you want to create the event log on the local Orion NPM server or a different server, next type the trigger message, then enter the alert reset message.

If you have an external syslog server, such as a Linux machine running the syslog daemon (`syslogd`) or an enterprise logging server such as Splunk, you can send a syslog message to that server. Alternatively, you can send an SNMP Trap message.

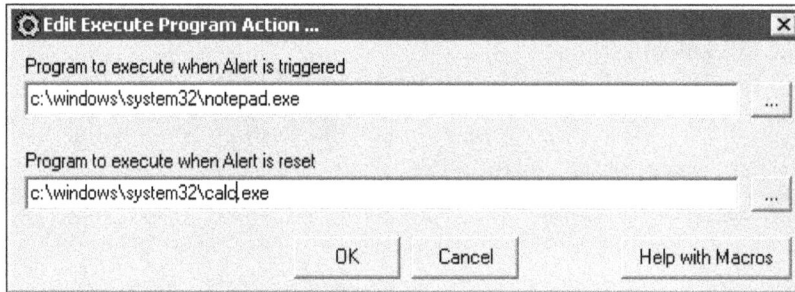

The two options, **Execute an external program** and **Execute an external VB script**, run a program or kick off a VBScript file as an alert.

The rest of the options are self-explanatory. **E-mail a Web Page** will send an HTML web page attachment to an e-mail address via SMTP, **Change Custom Property** will change a custom value on a monitored node, **Text to Speech output** will make the computer "talk" (which requires that sound be enabled on the server), **Dial a Paging or SMS service**, and **Get or Post a URL to a Web Server**. The **Send a Windows Net Message** option is something that is deprecated in Windows Server 2008 and above, so it is not recommended to use this feature whatsoever. It remains in Orion NPM for compatibility reasons.

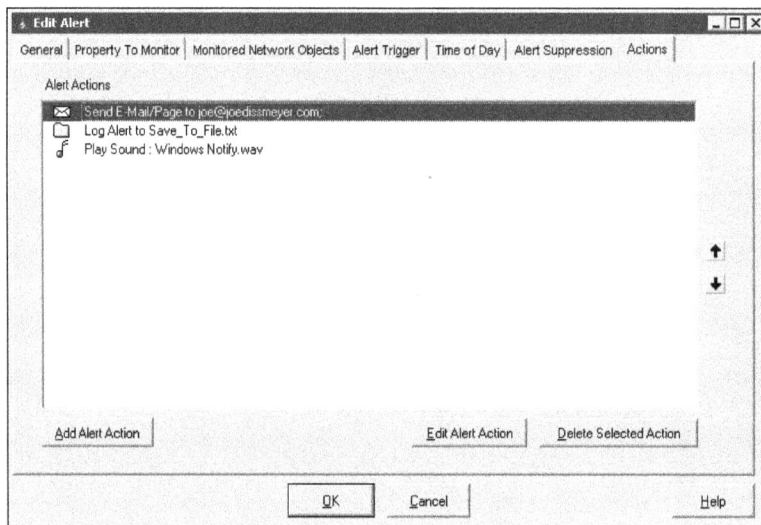

If you made a mistake, you can always delete or edit an alert in the list. Multiple actions can be taken per alert in Orion NPM, so feel free to add as many actions as you need. When finished editing the alert, click on **OK** to save it. The new alert will appear in the **Configure Alerts** window.

Configuring advanced alerts

Creating a basic alert does not take a great deal of time, but a basic alert is very limited. What do you do when you need to create an alert that will trigger only after a specific interface threshold has been breached? Or, what do you do when you need to set up an alert against a universal device poller? I will tell you what you need to do! Use **Advanced Alert Manager**!

Building an advanced alert is very similar to the process for building basic alerts. All advanced alerts are managed from the Advanced Alert Manager application installed on the Orion NPM server.

To launch the Advanced Alert Manager application perform the following steps:

1. Log into the SolarWinds Orion NPM server.

2. Navigate through **Start | All Programs | SolarWinds Orion | Alerting, Reporting, and Mapping | Launch Advanced Alert Manager**.

   Just like in the **Basic Alert Manager**, click on the **Configure Alerts** button to get started.

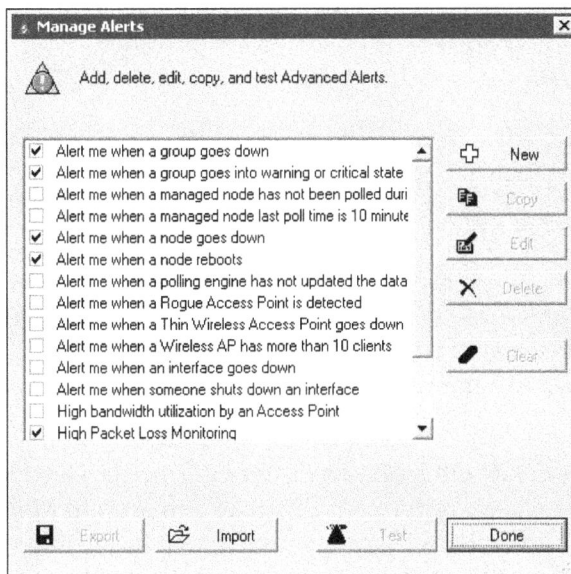

Right off the bat, you can see a major difference from the **Basic Alert Manager**. There are a few more options available in this view. You can see all of the advanced alerts pre-configured when Orion NPM was first installed. These are the same alerts discussed at the beginning of this chapter. Any alert that is enabled has a check mark next to it, while any disabled alerts are not checked. Two additional options are **Export** and **Import**.

You can export any alert definition from Orion NPM to a file. A saved alert file has the file extension as .AlertDefinition and can be imported on any other Orion NPM server. This feature is more or less used for sharing alert files within an organization. However, you can download and share alert definitions in the SolarWinds Thwack community as well!

Creating an advanced alert runs through the same process as creating a basic alert. Click on the **New Alert** button to get started. There is a new option in the **General** tab called **Alert Evaluation Frequency**.

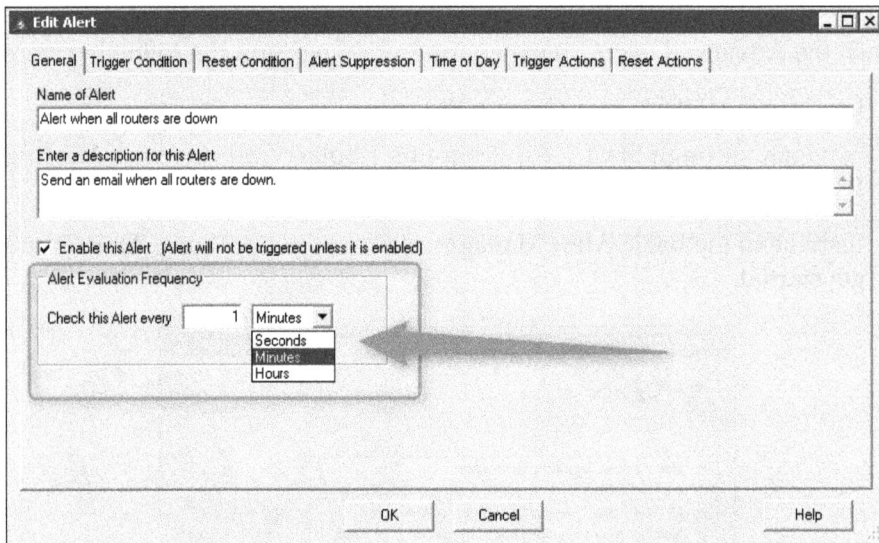

Alert Evaluation Frequency is a setting to decide how often Orion NPM is to check against the status of this alert. What this means is that depending on the frequency defined here, a duplicate alert will be generated until it is reset. For example, if you have the frequency set to 1 minute, the alert will trigger an alert action every minute until it is reset.

The **Trigger Condition** tab will look very different from the same tab in basic alerts. The one option most overlooked is **Type of Property to Monitor** at the top of the window. You can define a trigger based on a node, a volume, polling engine, interface, custom node poller, wireless access point, and more.

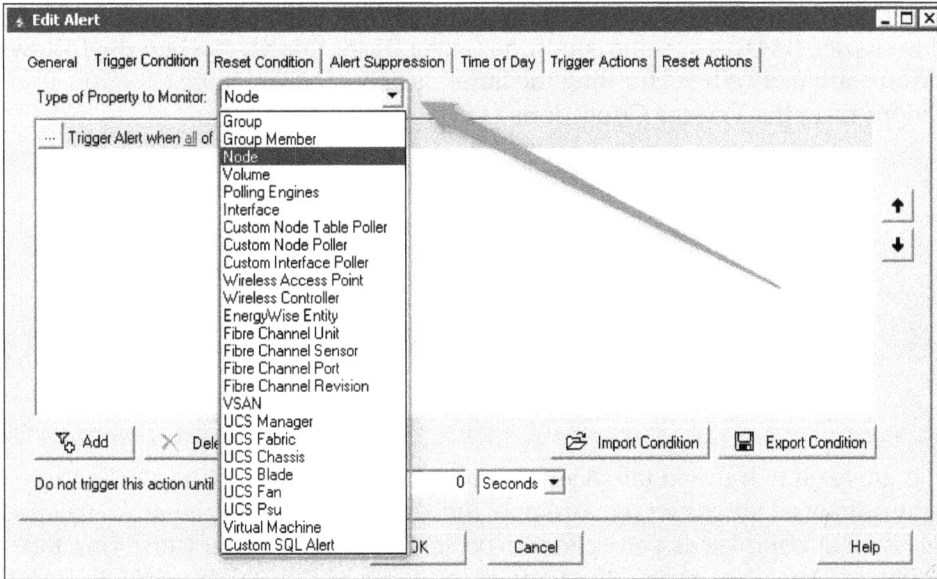

Once you choose a property type, click on the **Add** button at the bottom of the page and choose a condition to add. In a complex condition, you can change a variety of options, whereas in a basic alert you cannot. The following screenshot gives an example of how to define a complex condition:

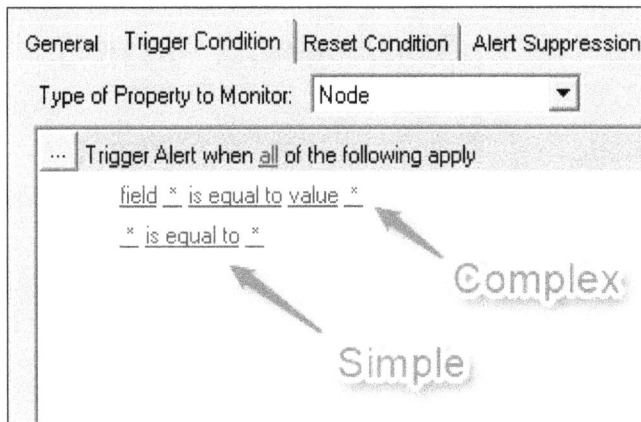

A complex condition is just like a math problem. Define what you need in order to get the results you expect. Click on the **Reset Condition** tab to continue.

Reset when trigger conditions are no longer true is the default option. But if you need to change this to a specific condition, select **Reset this alert when the following conditions are met**. When choosing the latter option, the same type of complex conditions from the **Trigger Conditions** tab can be selected here.

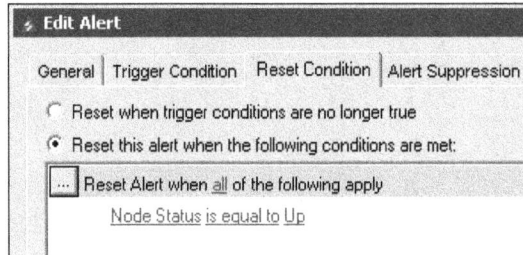

As you move on to the next tab, **Alert Suppression**, I'm sure you are starting to see why advanced alerts can be extremely time consuming! You can make the alert suppression as complex as you need it to be, just as in the **Trigger Condition** tab. The **Time of Day** tab is exactly the same as the **Time of Day** tab from the **Basic Alert Manager** where you can select the time and/or day of when to enable the alert.

Creating a new action in the **Trigger Actions** tab has the exact same options that the basic alerts has, but with a few special additions that I will address shortly. You can export selected actions to a .ALERTACTIONS file to be imported on another Orion NPM server and you can copy an existing action to the **Reset Actions** tab for quick editing.

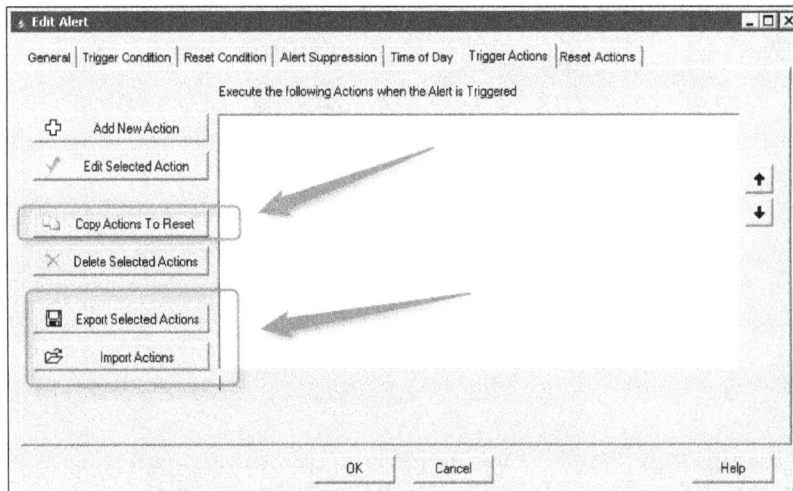

To display some of the additional options when defining an action in **Advanced Alert Manager**, click on the **Add New Action** button and choose **Send an E-Mail/Page**. The **E-Mail/Pager Addresses** tab is the same, but the **Message** tab has an **Insert Variable** button.

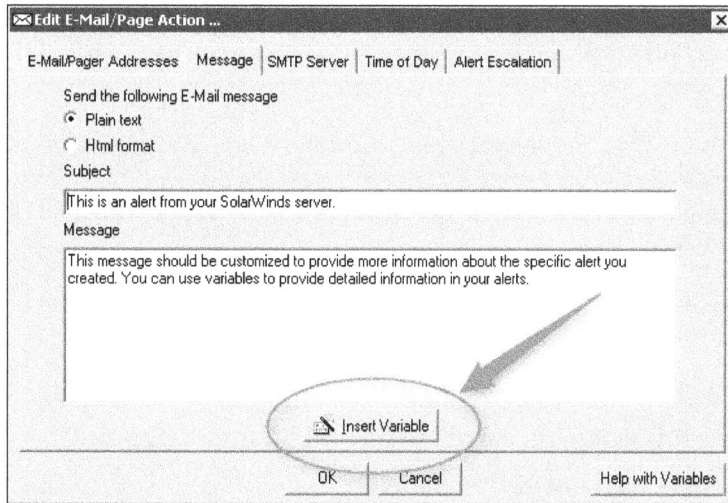

This allows you to find and insert a custom SQL variable from the Orion NPM database in the text of the message. For example, if you want to add the node's name to the e-mail message, click on the **Insert Variable** button then choose **General** under **Variable Category**, **NodeName** under **Select a Variable**, and click on the **Build Selected Variable** button.

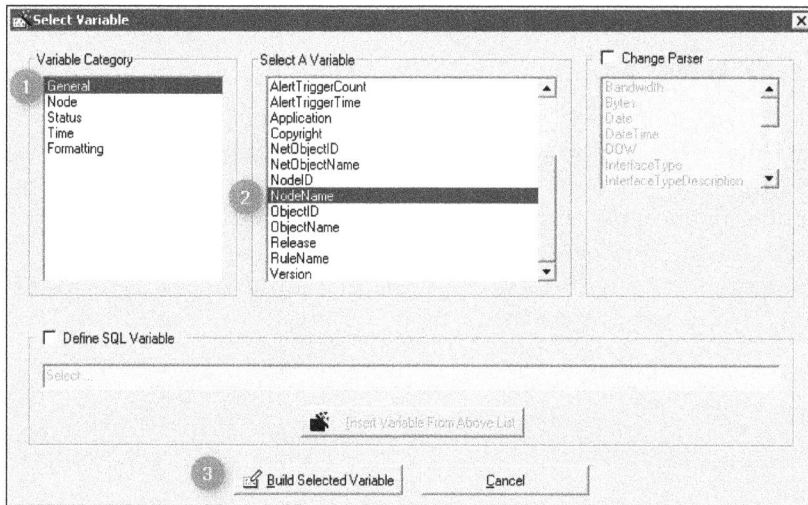

The variable will be added to the notification as defined by the dollar sign ($).

```
Subject
Alert! Node name - ${NodeName} - is down!
Message
The node, ${NodeName}, is down at the moment. Go get some coffee then fix this now!
```

These variables are SQL Server variables and can be heavily customized. If you are a SQL script junkie, then you will feel right at home! If you need help with building a variable, clicking on the **Help with Variables** button will open the SolarWinds Orion NPM web knowledgebase.

Define the **SMTP Server** and **Time of Day** settings, then click on **Alert Escalation**. This is a new option. There are three options available in this tab:

- **Do not execute this Action if the Alert had been Acknowledged**
- **Execute this action repeatedly while the Alert is Triggered**
- **Delay the execution of this Action**

These actions allow administrators the time to take action on a problem that they may already know about before an alert is generated, a trigger is hit, and an action is executed by Orion NPM. For example, an administrator may already know about a server that has gone down and just need some time to log into the Orion dashboard to acknowledge an event, which will prevent an e-mail notification from being sent out.

Testing alerts

Once an alert is built, it can be tested. To test an alert, click on the **Test Alerts** button in the **Basic Alert Manager** or **Advanced Alert Manager** to open the **Test Fire Alerts** window. First, you need to choose what you want to test fire the alert against. Choose a node, an interface on a node, or a volume. Second, choose the alert you wish to test. Last, click on the **Test Alert Trigger** button to test fire the alert. If there were any issues or errors with a test fire, click on the **View Alert Error Log** button for information on why the alert failed.

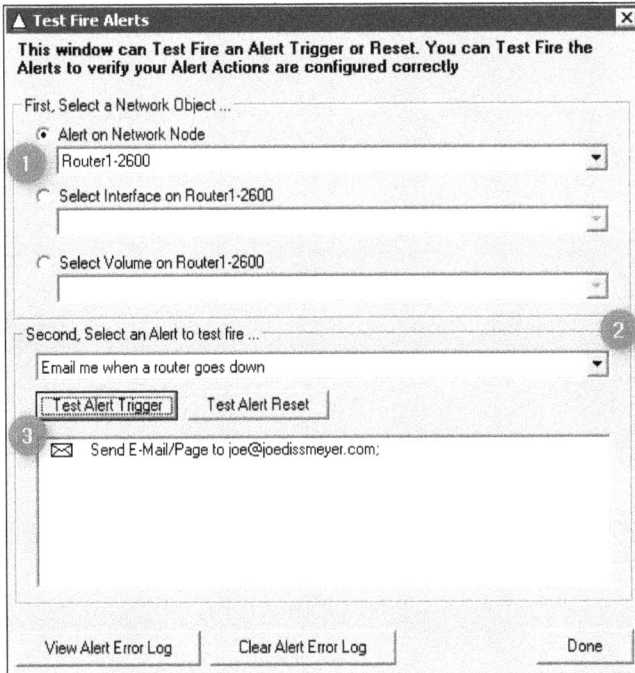

Summary

This marks the end of the chapter. In it, we discussed how the alert system in Orion NPM operates, as well as the basics of building both basic and advanced alerts. I have personally spent a great deal of time configuring some insane alerts because my customers' network environments required it. I'm sure you can see how important configuring alerts for your own network environment can be. Alerts can be extremely powerful, resourceful, and lead you in the right direction in the event of a wide-spread problem. In the next chapter, we will discuss how to produce reports, and network mapping.

7
Producing Reports and Network Mapping

My employer has several supervisors, managers, and directors that everyone must report to. The one thing that is especially true about managers is that they love reports! What is a report anyway? I like to think that a report is some type of historical information presented in an easy-to-understand format.

For network administrators, knowing information such as IP addresses, subnets, port counts, wireless access point signal levels, server CPU and Memory usage, and so forth is critical. We also need to know how traffic is entering our networks, and how traffic is exiting our network. SolarWinds Orion Network Performance Monitor can provide you with all of this information within the Reports feature.

In this chapter, we are going to fully discuss the **Reports** functionality in Orion NPM as well as its network mapping utility, **Orion Network Atlas**.

Understanding reports

At this point in time, you should be very familiar with how to monitor nodes in Orion NPM. In fact, I'm sure you must have noticed that there is plenty of historical data that Orion NPM saves in its database for every monitored object. Orion NPM allows you to take that data and export it into a report so that you can view its contents easily. Orion reports come in all shapes and sizes and the information that Orion reports can provide help the management and administrators with capacity planning, project management, vendor selection, and many other things.

You can run any of the pre-defined reports, which are installed with Orion NPM, on-demand within the Orion dashboard. Another thing to be discussed later is that you can schedule reports to be created and have them e-mailed to the appropriate people.

Every report available within Orion NPM can be found in the **Reports** view under the **HOME** tab in the Orion dashboard.

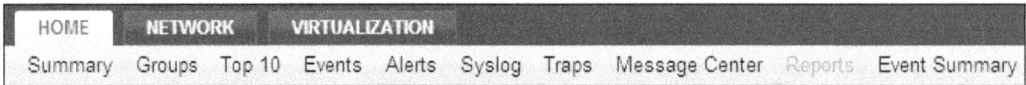

HOME	NETWORK	VIRTUALIZATION								
Summary	Groups	Top 10	Events	Alerts	Syslog	Traps	Message Center	Reports	Event Summary

The **Reports** page view will look similar to the following screenshot:

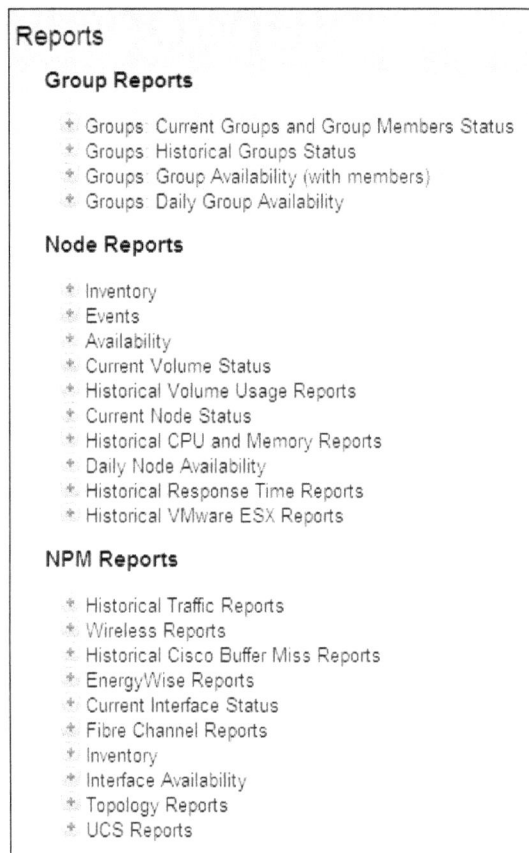

Reports

Group Reports

+ Groups: Current Groups and Group Members Status
+ Groups: Historical Groups Status
+ Groups: Group Availability (with members)
+ Groups: Daily Group Availability

Node Reports

+ Inventory
+ Events
+ Availability
+ Current Volume Status
+ Historical Volume Usage Reports
+ Current Node Status
+ Historical CPU and Memory Reports
+ Daily Node Availability
+ Historical Response Time Reports
+ Historical VMware ESX Reports

NPM Reports

+ Historical Traffic Reports
+ Wireless Reports
+ Historical Cisco Buffer Miss Reports
+ EnergyWise Reports
+ Current Interface Status
+ Fibre Channel Reports
+ Inventory
+ Interface Availability
+ Topology Reports
+ UCS Reports

As you can see, Orion NPM comes with several pre-defined reports, which are divided into the following three different groups:

- **Group Reports**
- **Node Reports**
- **NPM Reports**

Group reports are reports that are based on data gathered from a collection of monitored nodes. For example, I may have a group titled *routers* that includes all of my company's routers that I am monitoring with Orion NPM. Or, I may have another group titled *Customer #1* that includes all equipment that I am monitoring belonging to that single customer. Regardless of the group, I can run a variety of reports against a group such as the groups' availability or uptime, historical status of a group, or even a simple report such as a list of the groups' members.

Node reports are reports that you can run against information gathered from any of the monitored nodes or devices in Orion NPM. Node reports are not explicitly focused interfaces, but instead can provide inventory data, availability information, and various historical data on anything within a node itself. Using a node report, I could find out if my VMware ESX servers' CPU capacities are under-utilized, find out a total of how many stacked switches I have at a location, or find the average amount of storage used within a period of time. If there is a device that consistently reboots from time to time, a report can be generated against that node to find out exactly how many times Orion NPM created an event for it.

The final report type, **NPM reports**, are reports that can be made for any type of current or historical data within the Orion database and include reports for fibre channel, interfaces, volumes, wireless, network topology, and more. NPM reports can be thought as being a "catch all" for reports that may not fit within the other two categories. Using a NPM report, I can view a report that displays the current interface utilization for all monitored interfaces. I can also run a report against the historical data for each interface to find out what interfaces are under or oversubscribed. I can also run a specific report against Cisco switches based on EnergyWise information. The sky is the limit with NPM reports.

Running reports from the dashboard

There are two ways to run reports in Orion NPM. The primary way that you will execute a report is from the dashboard. From the **Reports** view, expand the report type you want to run and choose a report title.

As an example, I am going to run an inventory report that will display all of the various types of devices I am monitoring with Orion NPM. Under the **Node Reports** group, expand **Inventory** and then click on **Device Types**.

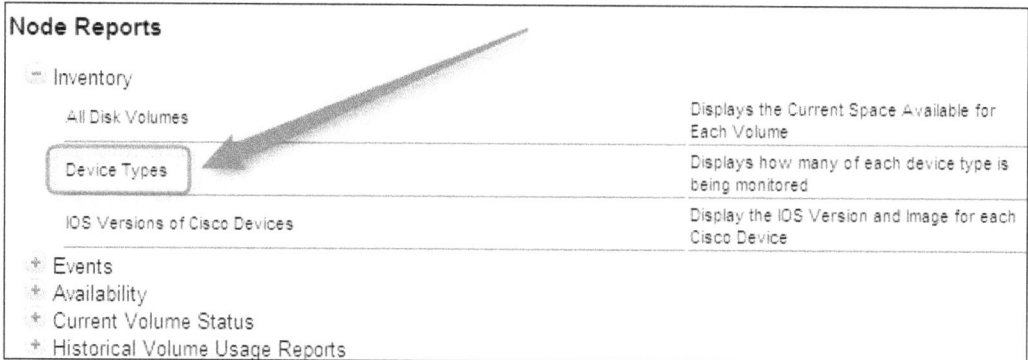

A new page view will display on the screen and will look similar to the following screenshot:

The **Device Types** inventory report is a very simple one. As you can see, I am monitoring various types of hardware from Windows Servers to Cisco access points, a VMware ESX server, and more. After a report is displayed, I can choose to print it or export it to a PDF file from the links on the upper-right corner of the screen.

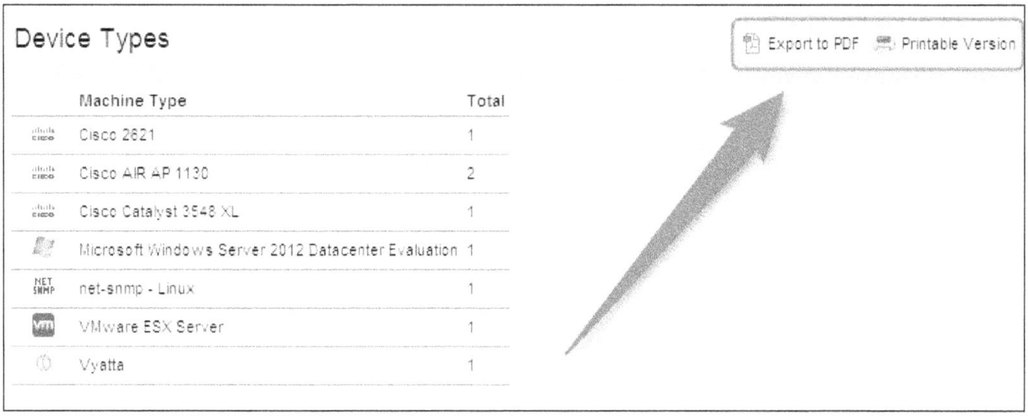

Choosing the **Export to PDF** option more or less creates a web page screenshot of the existing report inside a PDF file and will include all of the web page assets such as the menu bars, currently logged on user, and other data. An example of an exported PDF file is displayed in the following screenshot:

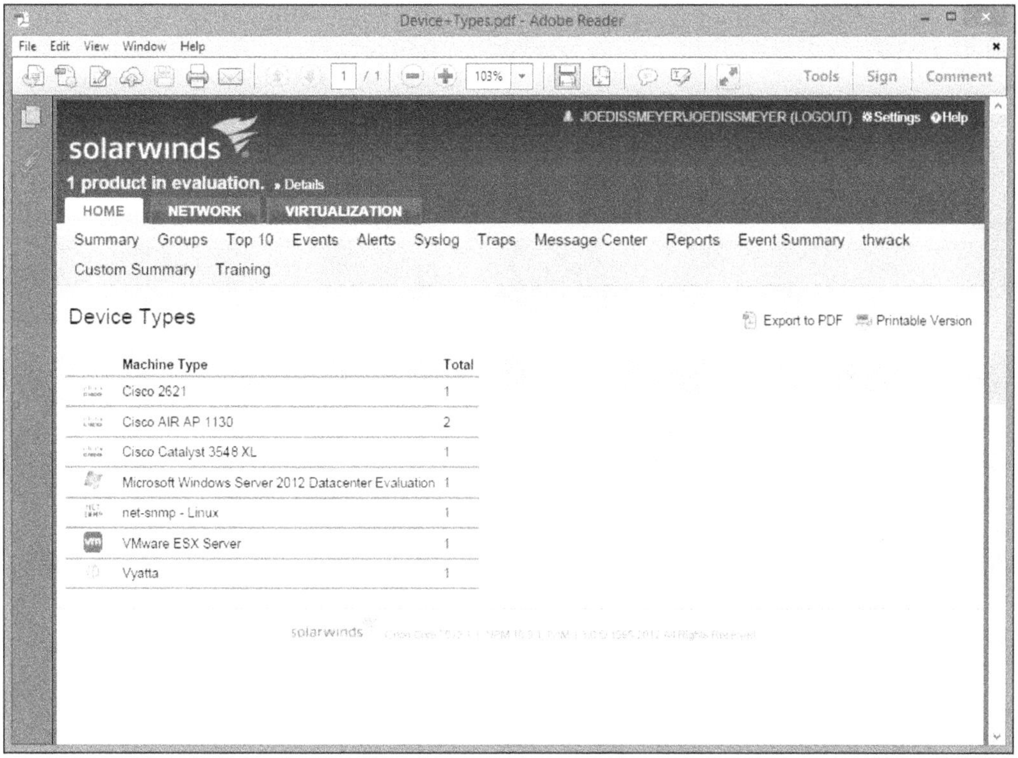

Choosing **Printable Version** will also create a web page screenshot of the existing report, but it will strip away all of the visual parts of the web page view (menu bars, SolarWinds title, and so on). An example of the printable version is displayed in the following screenshot:

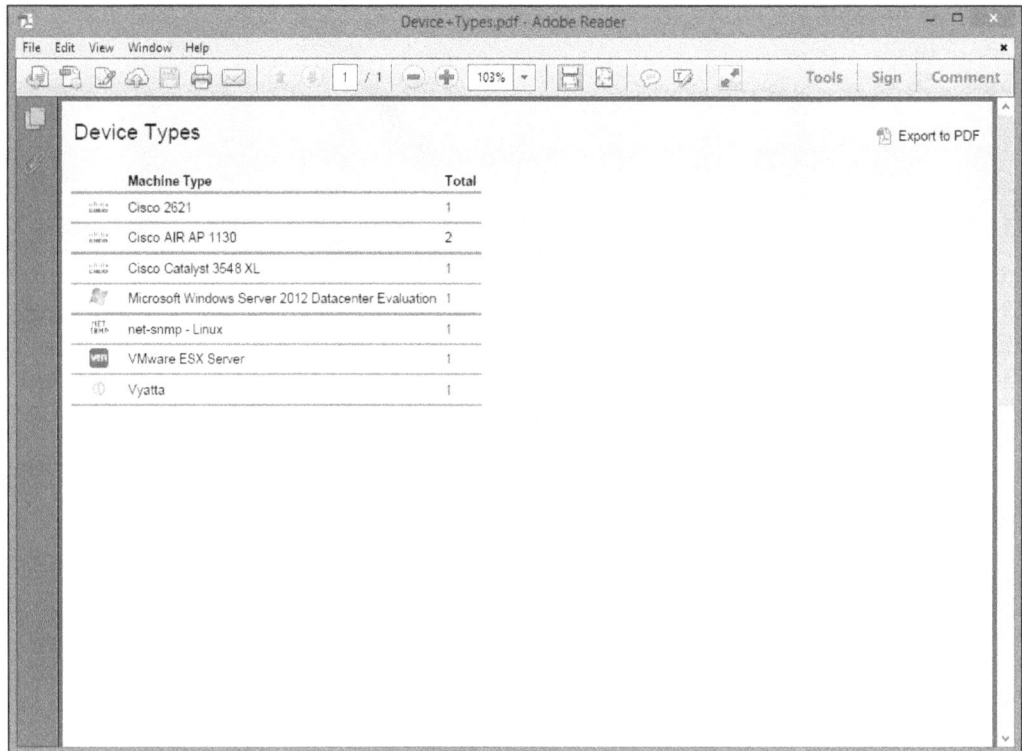

Running reports from Report Writer

The second, more advanced, way to run a report is from the Report Writer application. To launch Report Viewer, log onto the server where Orion NPM is installed on and click **Start | All Programs | SolarWinds Orion | Alerting, Reporting, and Mapping | Report Viewer**. Once the application is onscreen, click on a report title on the left-hand side pane, then click on the **Preview** button in the **Report Designer** window.

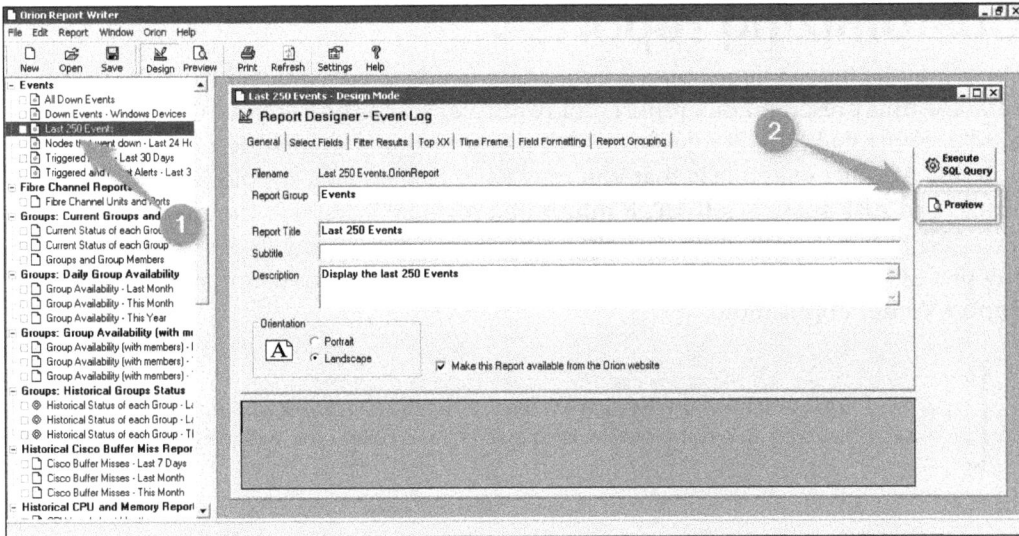

In the example, I scrolled down the left-hand side pane to the **Events** group and executed the **Last 250 Events** report. The report will look similar to the following screenshot:

These are only a few examples on how to run reports in Orion NPM. I encourage you to run several of the other built-in reports to see what SolarWinds has provided in an out-of -the-box installation. We will continue the discussion of how to create your own reports in the next section.

Customizing reports

There may be times where some specific data needs to be extracted from the Orion database and presented in a report. For example, you or one of your customers may need a report that lists all interfaces with a description that include the words *uplink* or *trunk*. Another example is that you may need to display a simple report on how much hard disk space is left on all monitored servers. Whatever the reasons may be, Orion NPM has the capability to help you produce almost any type of report from any piece of information in the Orion database. To create a new report, you use the **Report Writer** application.

> You must use the **Report Writer** application in order to create reports. It is not possible to create reports inside the web dashboard.

Creating new reports

When **Report Writer** is launched, it always displays the **Quick Help** view in the main display pane while the left-hand side pane lists every report (custom or built-in) that is available within the Orion NPM system. To create a new report, click on the **New** button in the toolbar or the **Create a New Report** button in the **Quick Help** view.

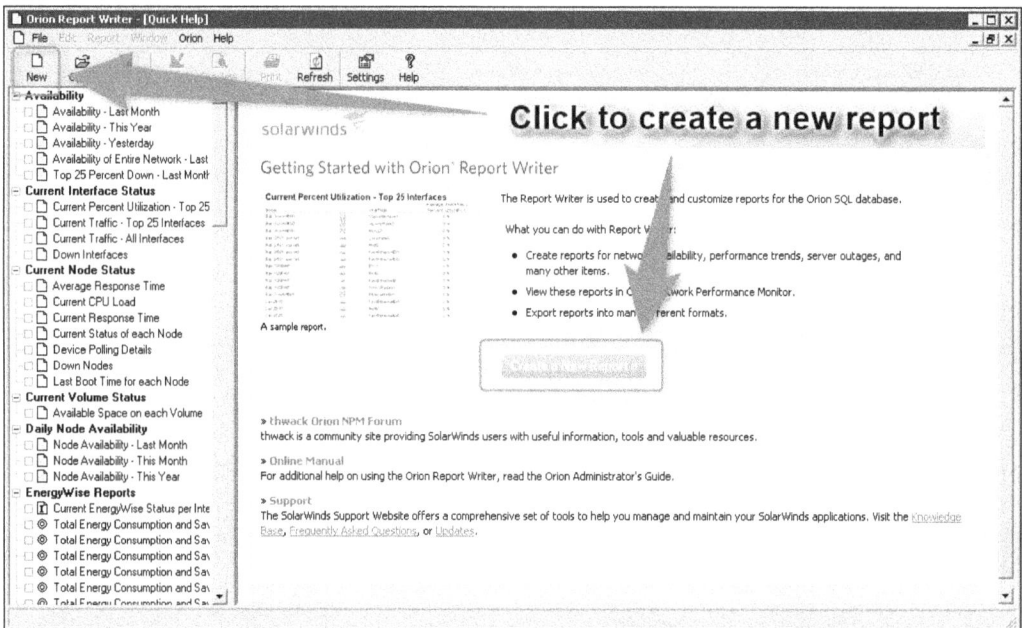

In the first step, you must make a decision of the type of report that you wish to create. There are several options available in the list and clicking one will display a short description of it.

For demonstration purposes, I am going to build a custom report that shows the current traffic of all interfaces with an administrative description of **Uplink** and/or **Trunk** on my switches. In the following screenshot, I chose **Current Status of Nodes, Interfaces, etc.,** then clicked on **OK** to continue.

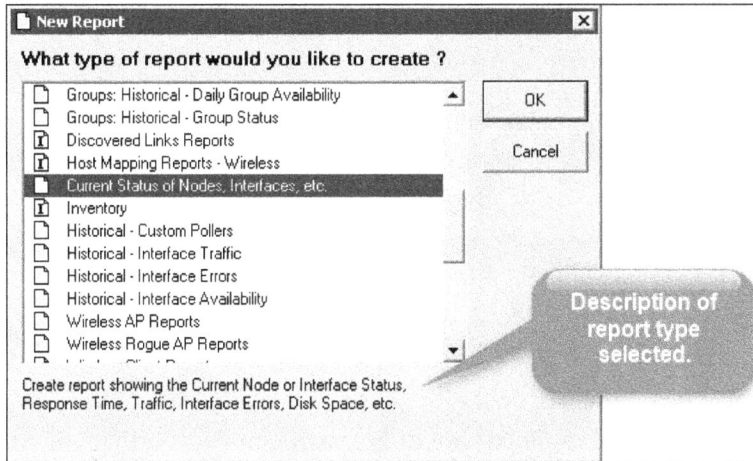

A new window appears in the **Report Writer** utility called **Report Designer**. The first tab is the **General** tab and an example is displayed in the following screenshot:

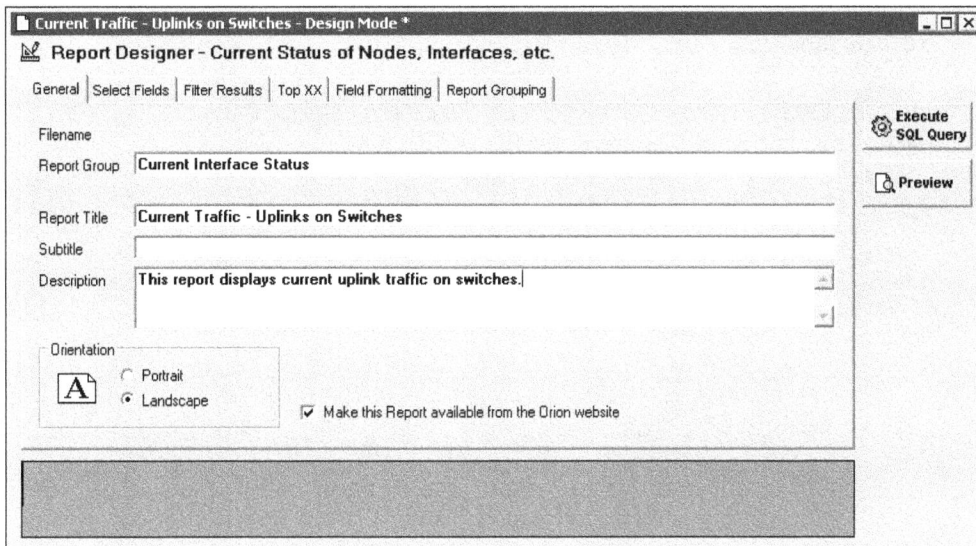

This is where you define several properties of the report itself such as the report grouping, the title of the report, and its description. You can also choose the orientation of the report whether it should be displayed in the portrait or landscape view, and you can choose whether or not to allow the report to be run from the Orion website. If you only want this report to be viewed from the **Report Writer** utility, then remove the check mark from **Make this Report available from the Orion website**. Click on the **Select Fields** tab to continue.

Select Fields is where you add in the data you want to view about the node. As shown in the following screenshot, click on the button next to the check mark and choose **Add a new field**:

From this point, you choose the field (**Node Name**, **Vendor Icon**, **Xmit bps**, and so on), how the field's data will be sorted (ascending or descending), as well as its function (sum, count, average, min, max). The default option for sorting and function is * (star) which stands for not defined or default. In the following example, I added all of the fields needed for my report, changed the sorting for the **Interface Caption** to **ascending**, and didn't define any functions for any of the fields. You will see why I chose these fields when the report is run, so please bear with me and continue to the **Filter Results** tab.

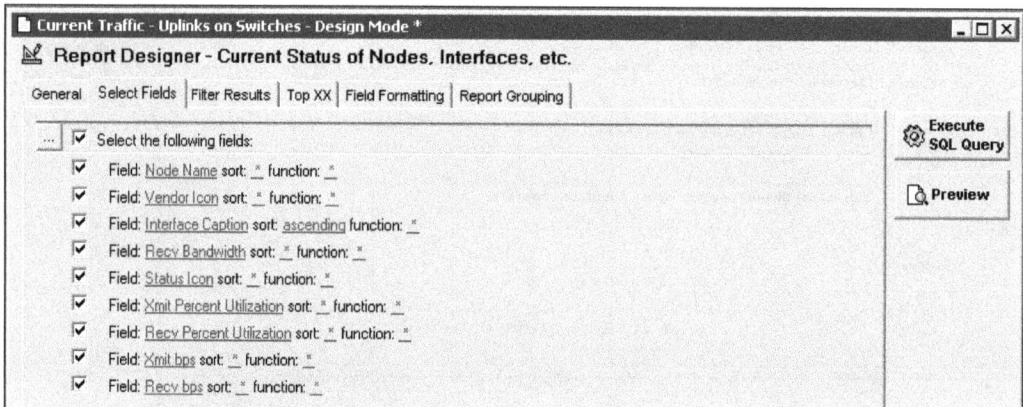

The **Filter Results** tab is where you limit, or specifically select, what data will be shown in the report. In the following screenshot, I added a filter to only include records where the **Machine Type** contains the word **Catalyst**, and only include interfaces that contain the words **Trunk** or **Uplink**. So when this report is executed, it will limit the data it displays to only my **Catalyst** switches and ports with the defined description.

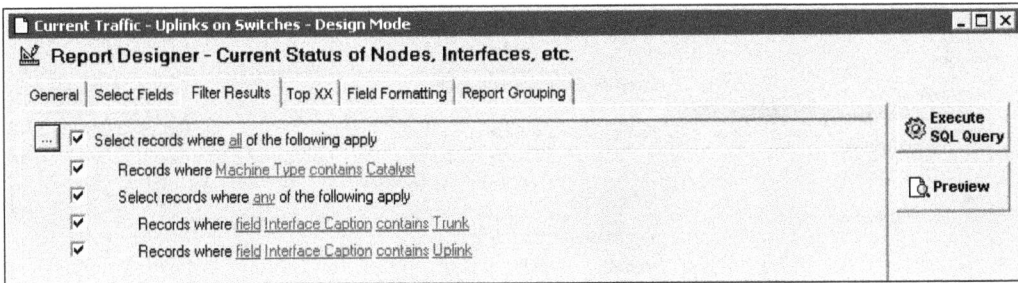

Next is the **Top XX** tab. Here you can choose to further limit the report views to only display a certain amount of records for the given report, such as the top 10 or top 5. For example, if you are building a report that needs to only display this information for the top 5 heavily-used interfaces across your network, then you will define such a value here. The default option is **Show All Records**, and is the option I will chose for my report.

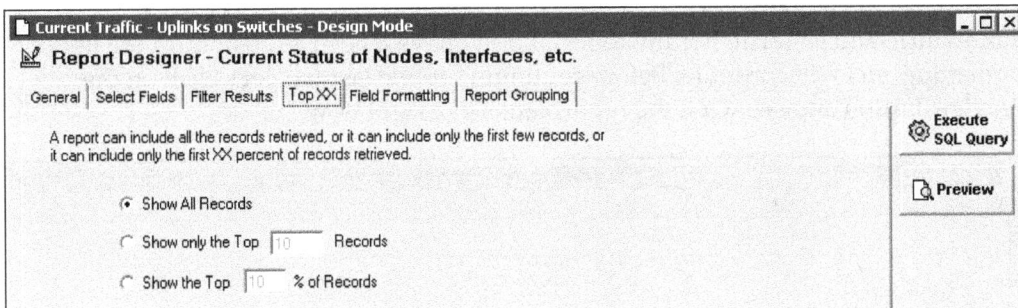

The next tab is **Field Formatting** and allows you to customize the formatting of each field as they are displayed in the report. An example of how to customize field formatting is displayed in the following screenshot:

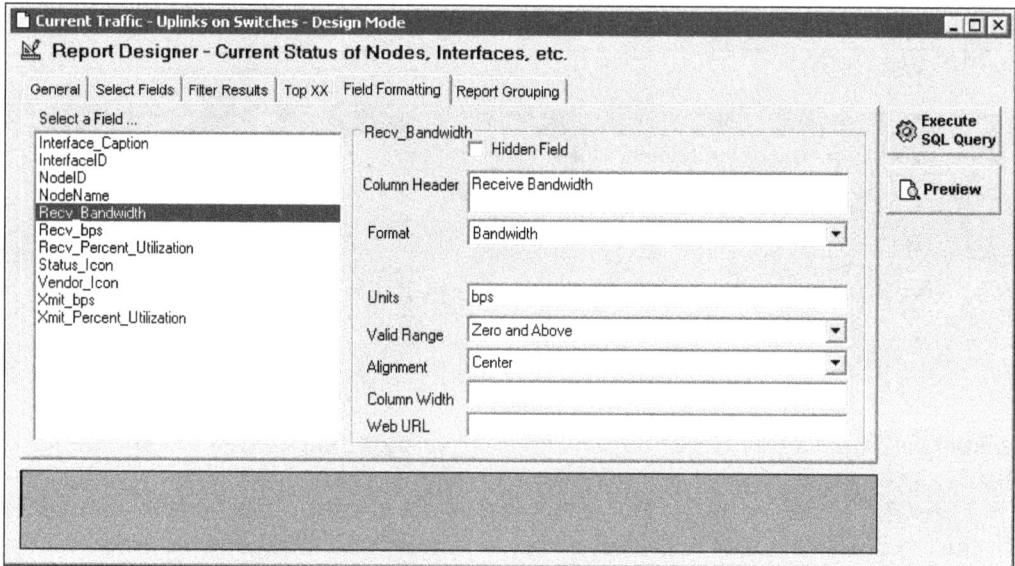

When clicking on the **Recv_Bandwidth** field, you can see what the column header's title will be, the numeric format of the data displayed which, in this case, is **Bandwidth**, and so forth. For my report, I don't have a need to customize the field's formatting, so I will continue. Before continuing to the last tab, let's click on the **Preview** button and see what the report looks like right now.

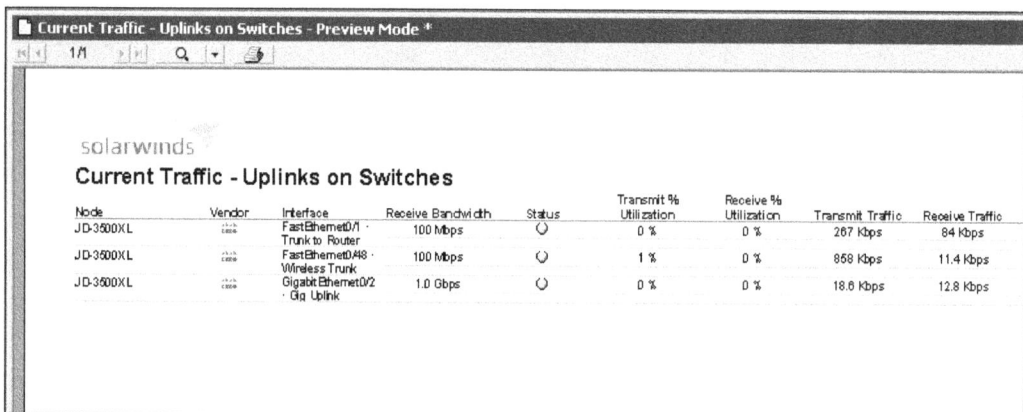

As you can see from the screenshot, this report is displaying everything I set out to do in this example. The report displays all fields in the order as shown in the Field Selection tab. But, what if I want to change how the report is viewed by grouping together similar items in order, such as grouping all interfaces that are "OK" (that is, have a green status light), or group all items from the same vendor in order? This is where you will use the options in the final tab, **Report Grouping**.

Report Grouping allows you to group similar, or "like", results together in the report. If I wanted to have my report, **Current Traffic – Uplinks on Switches**, display interfaces on the same node name first, then display the interfaces on another switch in the alphabetical order, I would add the **NodeName** field in the group as shown in the following screenshot:

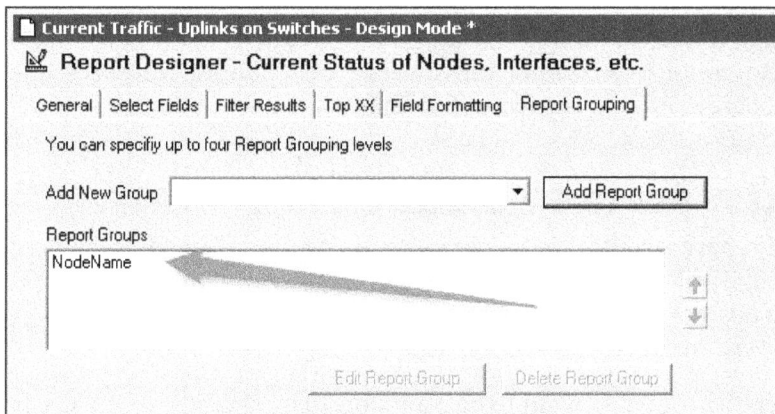

When clicking on the **Preview** button, you will see this rule in action in the next screenshot and it will group all interfaces together under the node's name.

If I had more switches in the list, the interfaces would be grouped under the node that each belongs to. So as you can see, there are almost an infinite number of ways to have Orion NPM display data on any of your monitored nodes.

> For more information on how to use the **Report Writer** application, as well as for information on how to create custom SQL queries, please see the Understanding Orion Report Writer documentation at `http://www.solarwinds.com/documentation/Orion/docs/UnderstandingOrionReportWriter.pdf`.

Once you are satisfied with your report, click on the **Save** button in the application toolbar. The report will now appear in its group on the left-hand side view pane in **Report Writer**, as well as its appropriate group in the Orion dashboard. The final task I want to complete is to embed my new report example in a page view in the Orion dashboard. The next section will demonstrate how to do this.

Adding reports to a page view

Any type of report can be added to an Orion dashboard page view such as the **Home Summary** page or **NPM Summary** page. This is a very useful feature, especially if you create a custom page view for a customer or department. To add a report to a page view, perform the following steps:

1. Open the page you want to add a report to by clicking on the **Customize Page** link on the upper-right corner of the view.

2. Click on the green plus button to add a resource to a column. In this example, I will add a resource to **Column 2**.

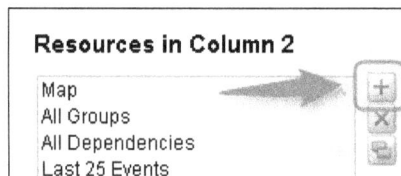

3. Expand the **Report Writer** group, place a check mark next to **Report from Orion Report Writer** and then click on **SUBMIT**.

Add Resources to Orion Summary Home Column 2

⊞ Node Lists - All Nodes and Grouped Node Lists
⊞ Summary Reports - Various Reports Showing Problem Areas
⊞ Network Wide Summary Charts - Charts Showing Statistics Across All Managed Devices
⊞ EnergyWise Controls - Charts and Reports for EnergyWise
⊞ EnergyWise Charts - Charts for EnergyWise
⊞ Top XX Lists - Top Response Time, CPU Load, Packet Loss, Traffic, etc.
⊞ Alerts - Defined and Triggered Alerts
⊞ Events - Event Summary and Detail Reports
⊞ Syslog - Syslog Summary and Detail Reports
⊞ Traps - Traps Summary and Detail Reports
⊞ Multiple Series Charts - Multiple Objects Chart & Multiple Universal Device Pollers Chart for a Single Node or Summary Page
⊞ Network Maps - Network Maps, Nodes on Maps, Lists of Maps, etc.
⊞ Inventory - Various Network Inventory Reports
⊟ Report Writer - Turn a Report from Report Writer into a Web Resource
    ☑ Report from Orion Report Writer
    ☐ Custom List of Reports
⊞ thwack - thwack Resources
⊞ Miscellaneous - Miscellaneous Resources
⊞ VSAN Summary - VSAN Summary Resources
⊞ What's New - What's New in Orion NPM 10.3
⊞ Virtualization Summary Reports - Resources that display overall application status

SUBMIT

---

4. Reorder the resources in the column as you see fit and click on **DONE**.

5. Click on the **EDIT** button in the **Report from Orion Report Writer** resource on the web page.

Report from Orion Report Writer   EDIT | HELP

A report has not been selected for this resource.
- **Select a report** for this resource.
- Click **Edit** in the resource header at any time to configure this resource further.

» Configure this resource

---

6. Choose the report you wish to display from the drop-down list then click on **SUBMIT**. In the example, I am going to display my custom **Current Traffic – Uplinks on Switches** report. Orion NPM will automatically display the correct report title, so there is no need to customize the resource title.

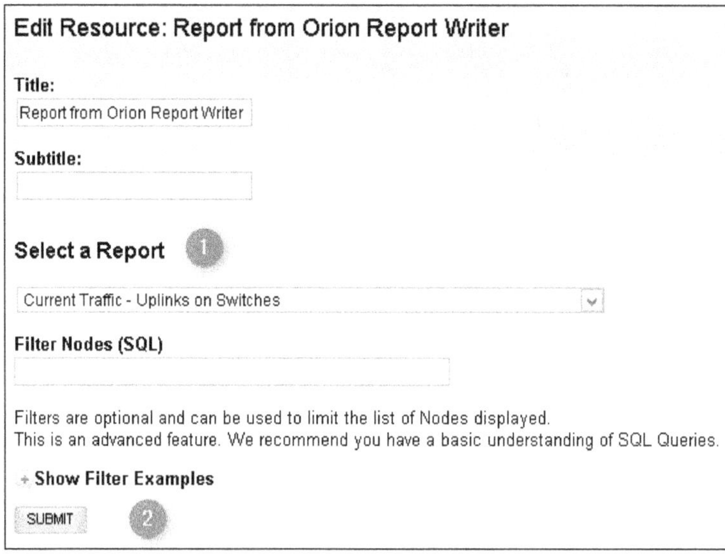

The report will now display inside of the web view as shown in the following screenshot:

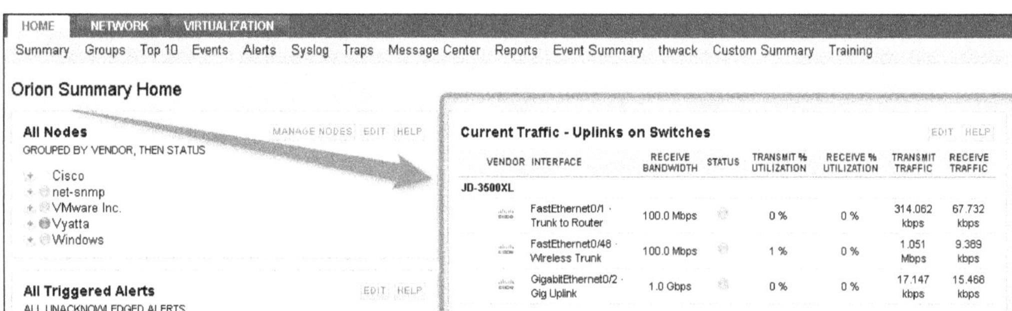

With this example, you have seen how you can add your own reports to web views in the dashboard. In the next section, we will discuss how to schedule execution of reports and how to automatically e-mail them to a recipient.

Working with the Report Scheduler

The Report Scheduler is used to automatically run reports and schedule Orion NPM to perform an action when that report has completed. You can e-mail a report in HTML format, e-mail a report in PDF format, or automatically print the report to a network printer. The most common action one would take is to schedule a report to run and have it e-mailed to a person, in PDF format. If you need to regularly send network reports to a manager or customer, use the Report Scheduler to do the job on your behalf!

You will find the application in **Start | All Programs | SolarWinds Orion | Alerting, Reporting, and Mapping | Report Writer**. After launching it, a basic start screen is displayed. By default, there are no scheduled reports after Orion NPM is installed for the first time.

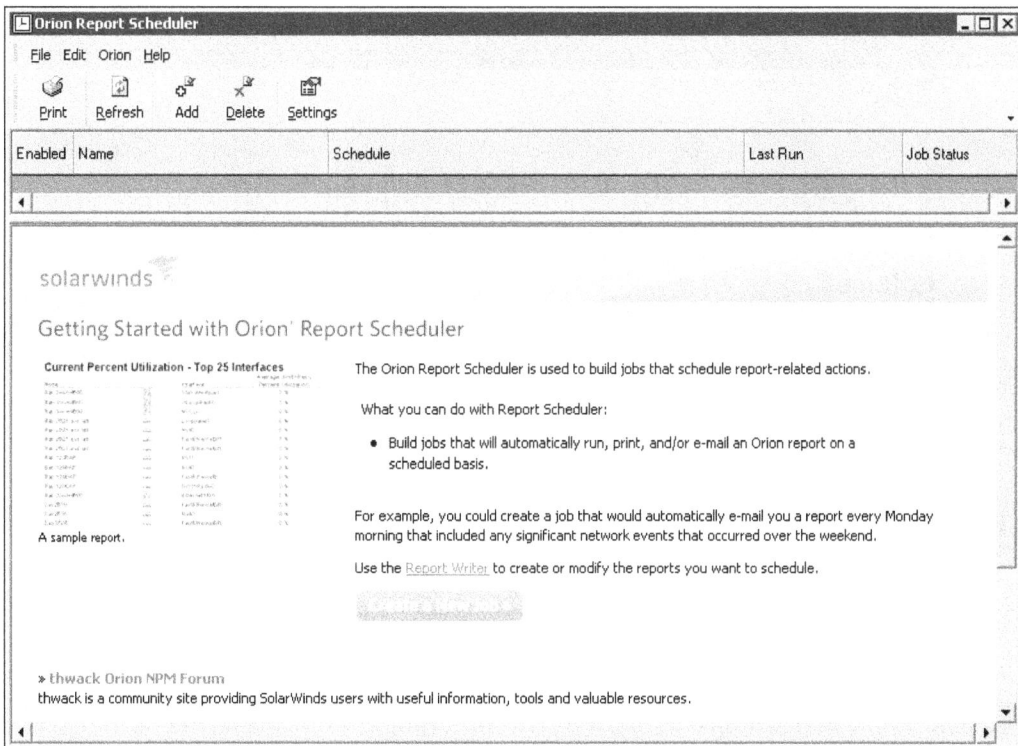

One of the first things that you should do when launching the **Report Scheduler** is to define the program's settings. In the toolbar, click on the **Settings** button.

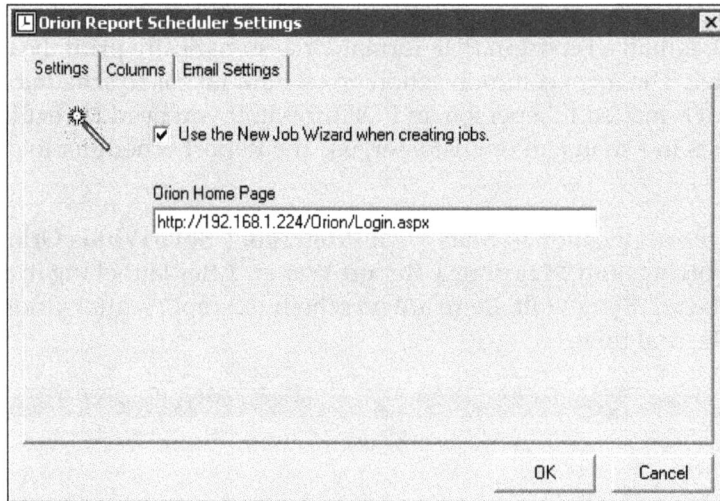

There the following are three tabs in the **Settings** menu for the **Report Scheduler**:

* **Settings**
* **Columns**
* **Email Settings**

The **Settings** tab is where you enable the option to use the New Job Wizard when creating jobs, and is where you define the Orion home page URL. The New Job Wizard makes it easy to create a scheduled job, so I recommend leaving the New Job Wizard option checked. When you create a new job, you have to define the URL in the Orion dashboard from where the report is run.

> If you don't know the exact URL of a report, just open a browser, log in to the dashboard, and copy/paste the report URL from the address bar.

The **Columns** tab is where you select the information to be displayed about each scheduled job. The fields defined in this option are the tables that appear only in the **Report Scheduler** and help to display more information about that job. Use the default options in this tab unless you want to remove or display more information about the scheduled job.

The **Email Settings** tab is where you define the SMTP e-mail settings for your organization. You will want to have these settings ready to go before you start scheduling jobs.

Creating a new job is very easy. Just click on the create new scheduled job button on the main page, or click on the **Add** button in the toolbar of the **Report Scheduler**.

First, you need to give the scheduled job a descriptive name. Make sure this is a meaningful name and click on the **Continue** button to go to the next step.

In step two, you need to define the URL of the report in the Orion dashboard. To do this, perform the following steps:

1. Click on the browse button on the right-hand side side of the window.

2. A web browser window will display the Orion dashboard. Click on the **Reports** link under the **HOME** tab in the menu bar and choose a report to display.

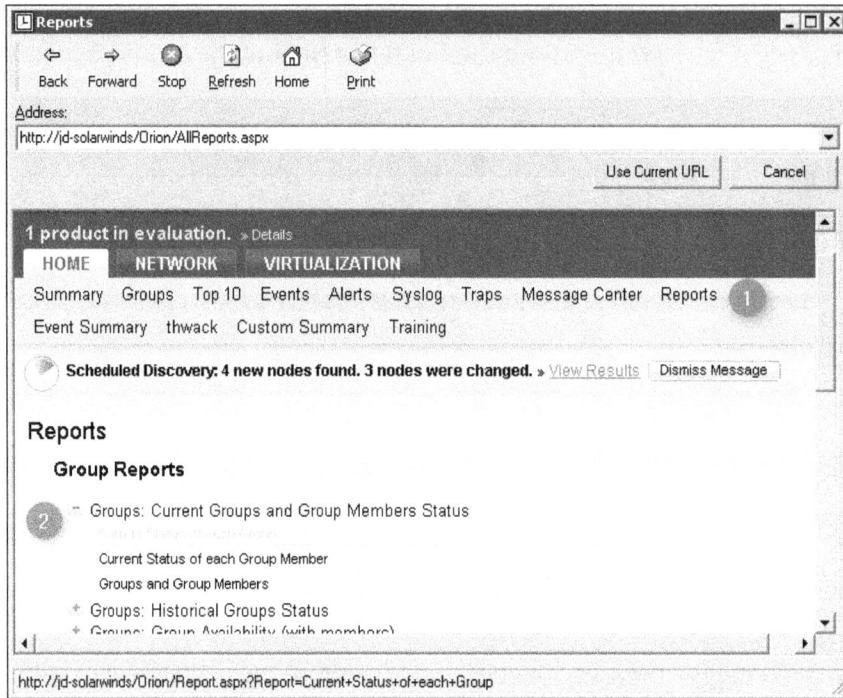

> A login prompt may appear when the browser window appears. If you have issues attempting to log into the dashboard, open the **Orion Report Scheduler Settings** and change the Orion Home Page to `http://LOCALHOST/Orion/Login.aspx` instead of the server's IP address or DNS host name, then try again.

3. Once you have chosen the report, click on the **Use Current URL** button to select it. Alternatively, you can type in the exact URL of the report if it is known.

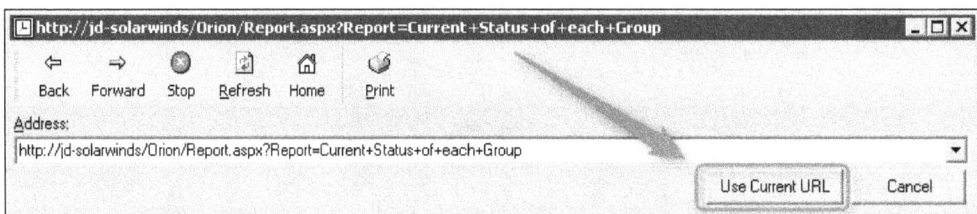

4. On the **Orion Web Login** tab, choose whether or not you want to produce a scheduled report in the Orion dashboard "printable format" and if you want Orion NPM to send the username and password in the e-mailed report. If you created a basic user account that has rights to the reports in the Orion dashboard, but the user receiving the report does not have access to Orion, you may want to include the login credentials in the e-mailed report.

5. Click on the **NT Account Login** tab and define the user account that has access to view this report only if Orion NPM is configured to accept Windows Active Directory account logins.

6. Click on **Continue** to move on to the next step.

Step three is where you define the report's automatic schedule. Choose the frequency of when the report needs to run, the date, time, and day. Click on the **Continue** button to move to the next step.

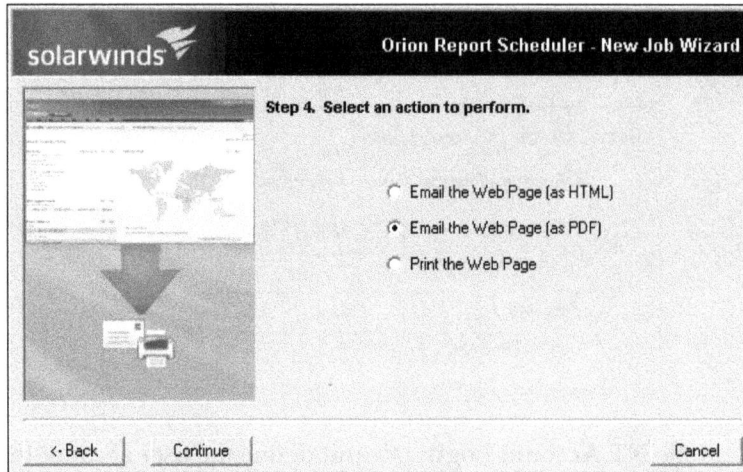

In step four, you need to choose how you want the report to be delivered to the user. There are only three options available, e-mail a copy of the report in HTML format, e-mail a copy of the report in PDF format, or print the report to a network printer. The most common option is to e-mail the report in PDF format to the user. Spam filters tend to block HTML e-mails more than those with PDF attachments, but the choice is up to you. If you choose to print the report to a printer, as soon as you click on **Continue**, you will need to choose the printer. The printer must be connected to the Windows Server that is running Orion NPM.

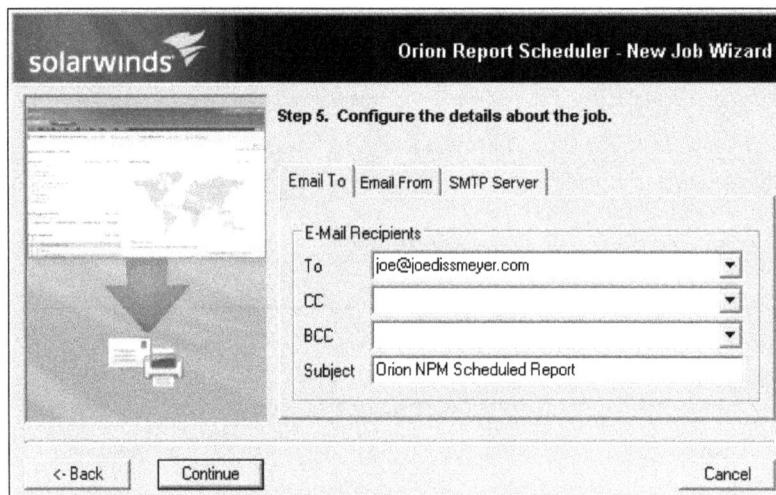

Define the e-mail addresses that should receive the report and enter some type of subject line. The **Email From** and **SMTP Server** tabs should already be pre-populated with the same SMTP information from the **Orion Report Scheduler Settings** discussed in the beginning of this section. If not, you may enter your organization's SMTP settings manually. Click on the **Continue** button to move to the final step.

In step six, define a user account that will have access to Orion NPM that will run the scheduled report. If you have a domain service account configured for the SolarWinds Orion NPM installation, it is recommended to use that account instead of an administrator's domain user account. If you are running Orion NPM on a workgroup server, use a local user account on the Windows Server. Click on **Next** to continue.

Finally, enter an administrative note or description for this job and click on the **Finish** button to save the job. The scheduled job will be saved and it will appear in the main display pane in the **Orion Report Scheduler** application.

You can always go back and edit an existing job, copy or duplicate a job, delete a job, or run the job on demand. Right-click on a job in the display pane to view the available options.

Network maps

It is true that a picture speaks a thousand words. It is much easier to troubleshoot a problem, or plan a deployment, or see what is working and what is not working when you have a real idea of what your network currently looks like. Orion NPM includes (at no additional charge!) a tool that allows you to build custom visual maps for your networks and apply them to your dashboard views. This tool is called the **Orion Network Atlas**.

Orion Network Atlas is a **what-you-see-is-what-you-get** (**WYSIWYG**) type of application that displays how the network map will literally look like in the Orion dashboard as you customize it. It is an extremely powerful application and a great compliment to the Orion NPM product. You can make as many maps as you like and apply any number of these maps to any view in Orion NPM. For example, you can create a map for one of your locations, create a custom view for that location in Orion NPM, then add that map to the custom view. Or, if you have multiple customers with multiple locations, you can do the same thing but on a much broader sense.

To download and install Orion Network Atlas to run on your workstation, click on the download link in the **Map** module on the Orion dashboard and run the EXE file. Otherwise, the application is installed by default on the Windows Server where Orion NPM is installed. You can find it in **Start | All Programs | SolarWinds Orion | Alerting, Reporting, and Mapping | Orion Network Atlas**.

If the download link for Network Atlas is not present, click on the **EDIT** button in the **Map** module and make sure there is a check mark next to the **Show Network Atlas Download Link** option. Otherwise, you can find the installer at the following local folder location on the SolarWinds Orion NPM server:

```
%systemdrive%\inetpub\SolarWinds\NetworkAtlas\
NetworkAtlas.exe
```

Choose all of the default options during the installation wizard. After the installation has completed, launch **Network Atlas** from the Start menu. Every time you launch **Network Atlas**, the **Connect to server** dialog box appears. A sample of what you may see is as follows:

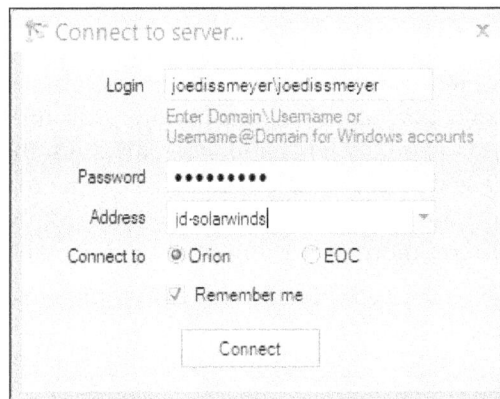

At the **Connect to server** dialog box, you must choose the following:

- **Login**
- **Password**
- **Address**
- **Connect to**
- **Remember me**

The **Login** and **Password** fields are where you enter the same credentials you use to access the Orion dashboard. The **Address** field is where you define either the DNS host name or IP address of the Orion NPM server. Finally, you can choose to have Windows remember these settings in your Windows user profile. This will make it easier to re-launch the application without having to type all of this information again. Once you get connected to the server and the Network Atlas appears on the screen, you can start building maps and apply them to the Orion dashboard views. Let's navigate the various display panes and options within the **Network Atlas** application itself.

Network Atlas overview

I must warn you that I can literally write a book solely on just creating network maps with the Orion Network Atlas tool! But I will do my best to be concise.

The two primary areas of interest when editing or creating maps are the object pane and the edit pane. The **object pane** is to the left-hand side and is where all of your discovered nodes are listed as well as your custom maps. While building a map, you drag-and-drop objects from the left-hand side pane onto the right-hand side edit pane. The **edit pane** is where you arrange objects on the map and is where you customize its visualization. You can change the background, add or remove objects, change symbols, and more. Anything saved in this pane will appear in your network map.

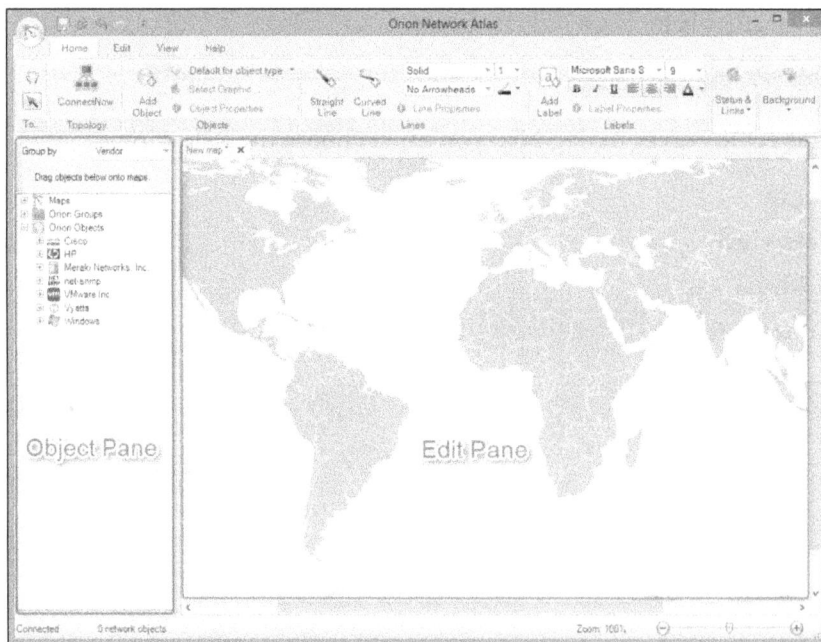

The ribbon tabs at the top of the window will help you with editing your maps. If you have ever used Microsoft Visio, or a similar visualization tool, then you should feel right at home.

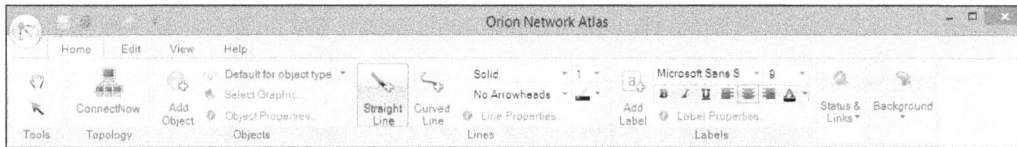

The **Home** tab is where most of the common editing tools are located. From left-to-right are the following menus:

- **Tools**
- **Topology**
- **Objects**
- **Lines**
- **Labels**
- **Status & Links**
- **Background**

There are two tools available in the **Home** tab; the Pan tool (hand icon) and the Select tool (the cursor icon). The Pan tool is used when moving around very large maps while the Select tool is used to manipulate objects in the edit pane.

ConnectNow is the only **Topology** tool and is used to automatically link devices together in the edit pane. For example, if you have CDP enabled on all of your routers and switches, Orion NPM will detect and record CDP neighbors. **ConnectNow** can make it easy to build complex network maps, provided that Orion NPM already knows about each device's relationship.

> In order for Orion NPM to use **ConnectNow** to automatically be able to link objects in Network Atlas, Orion NPM needs to have the link discovered during a network scan. If you do not have network discovery enabled, then there is a good chance that **ConnectNow** will not work.

The **Objects** section is where you can select an array of inanimate objects to add to your map, change the icon type of any objects added to the map, and edit any of the object's properties. For example, I can add a building icon to the map to indicate an ISP's central office on the map. To add this object, click on the **Add Object** button, then choose **Select Graphic** and choose a building icon.

The **Lines** section is where you define the various connectors and lines that link objects together. You can edit the line type (dash, dots, solid line), a line's color, straight or curved line, and more. An example of what you can make a line look like is shown in the following diagram:

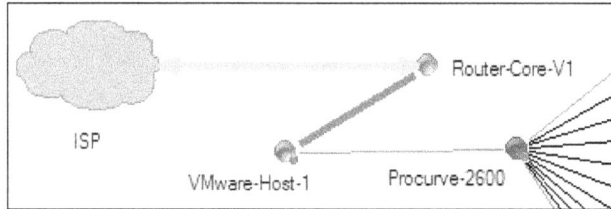

Labels is where you manage the appearance of an object's label. There is a full line of options for fonts, font sizes, font colors, and various appearance options for labels.

Status & Links is where you can modify the hyperlinks for an object in the network map. By default, when clicking on a node in a network map in the Orion dashboard, you will automatically navigate to that node's details view. Using **Status & Links**, you can edit those URL links to something different. Also, you can change what type of information the object displays in the map. A node's status (green or red status icon) is displayed by default and can be changed to a completely different appearance, such as the node's metrics. The options are virtually endless.

Background is the final option in the **Home** ribbon tab where you can change the background properties of a map to a different image, a linked image (via URL), the background color and/or texture.

Next up is the **Edit** ribbon tab.

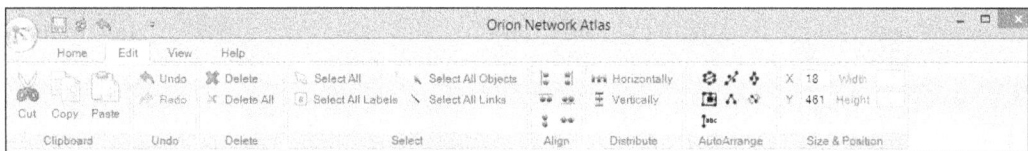

All of the options in the **Edit** tab are self-explanatory. However, the only item of interest are the **AutoArrange** options. If you are able to use **ConnectNow** to automatically link objects together, **AutoArrange** can automatically arrange the layout of the connected nodes for you with these options, instead of you needing to manually maneuver each node one-by-one in the **Edit** pane.

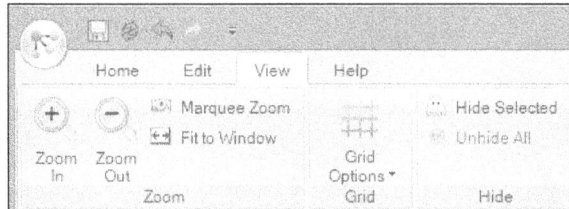

Every option in the **View** tab, as seen in the previous screenshot, is also self-explanatory. You can zoom in and out of a map, fit the map to the edit pane window, and choose various grid options. Enabling the grid may aid you in making sure objects are aligned correctly when moving around many objects in the edit pane.

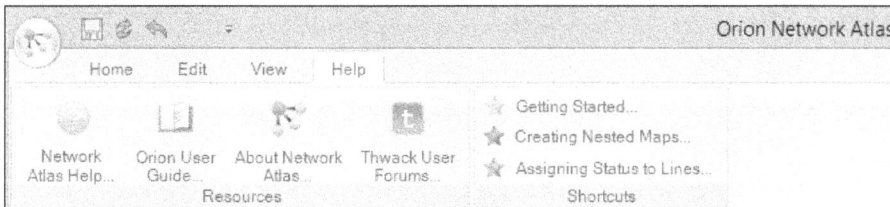

The final tab, **Help**, has resource and shortcut links to the Orion NPM online help guides, the Thwack communities, and user guides.

The **Getting Started** window in Network Atlas is a great starting point for learning how to build network maps and apply them to views in Orion NPM. It appears every single time you launch Orion Network Atlas, unless you disable it by clicking on the **Don't show again** option at the bottom of the window.

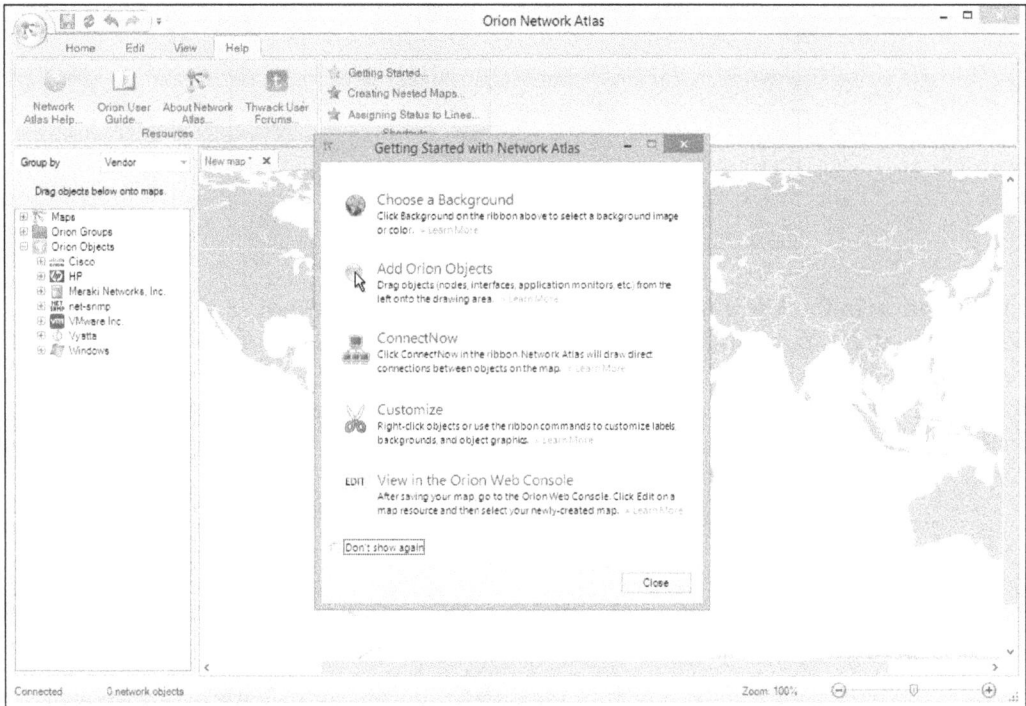

Building network maps

The following is a general workflow, from start to finish, when building maps in Orion NPM:

1. Choose a map background.
2. Add network objects to the map.
3. Use **ConnectNow** to automatically apply discovered relationships/connections.
4. Customize labels, graphics, and backgrounds.
5. Apply the map to a view in the Orion dashboard.

The entire point of building a network map is to have a visual representation of the network design, how a network has been engineered at a specific location, or a combination of both. Next, we will see how to build a basic map using the **Orion Network Atlas** application, then apply that map to the **Home Summary** view in Orion NPM.

Click on the application orb on the top-left corner of the **Orion Network Atlas** window. Choose **New Map**.

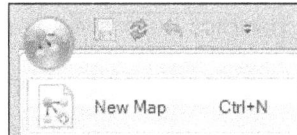

Choose a background image by clicking on the **Home** tab in the ribbon and then on the **Background** button. This is optional and can be changed later.

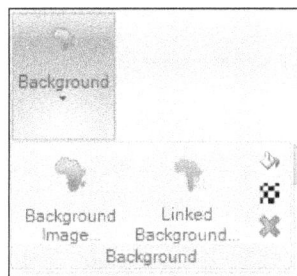

Drag-and-drop any, or all, objects from the object pane onto the edit pane. You can do this one by one, or simply drag an entire group.

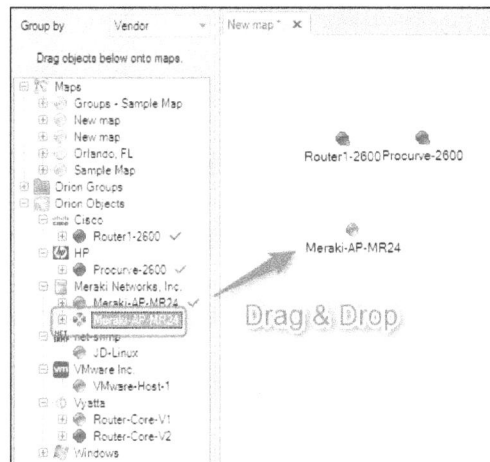

Notice that as you add objects to the edit pane, a green check mark appears next to the object's title in the object pane.

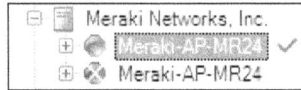

After you have added the appropriate network objects to the map, add additional representative objects as needed. A common object is a "cloud icon" to represent the Internet. To add such an object, perform the following steps:

1. Right-click on a blank space on the edit pane and choose **Add object**.
2. Right-click on that new object and choose **Select Graphic...**.

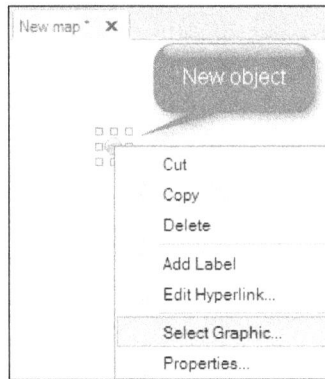

3. Under **Networking Icons**, highlight **WAN**. Choose a cloud icon and click on **OK**.

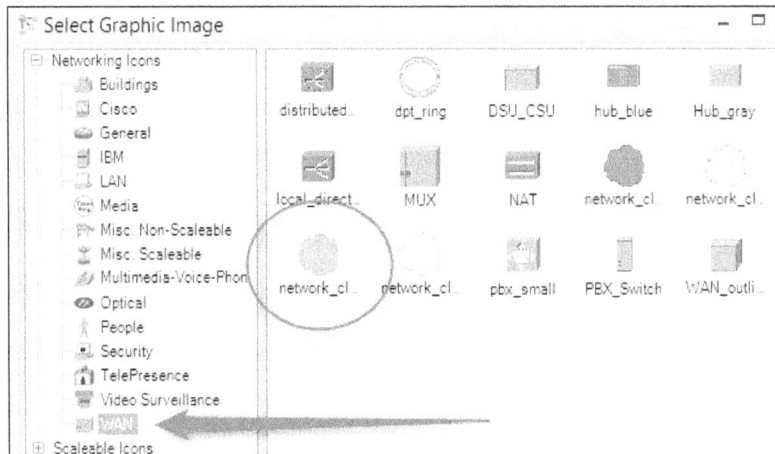

4. The new icon will appear in the edit pane. Add a label to the icon by right-clicking on it and choosing **Add Label**.

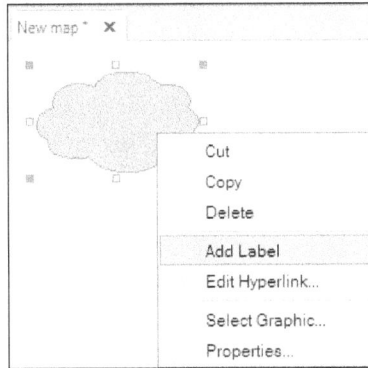

5. Double-click on the label to edit the text. Use the **Labels** section on the **Home** tab in the ribbon to customize the label's attributes.

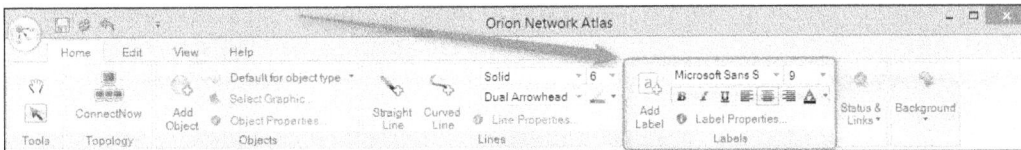

Now that some network objects are added to the edit pane, it is time to connect them together using lines. There are two ways to do this. The first is manually by using the lines from the **Home** tab in the ribbon. Choose a straight or curved line, then click-and-drag from one object to another. This will connect each object together, representing either a logical or physical link. Keep in mind that you can edit the appearance of a line from the **Lines** section in the **Home** tab. The following example shows that I created a line that links two of my monitored nodes with a thick blue line that includes bi-directional arrowheads.

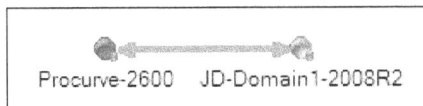

The second way to link objects is by using **ConnectNow**. Use the Select tool to select all objects that are monitored by Orion NPM and then click on the **ConnectNow** button.

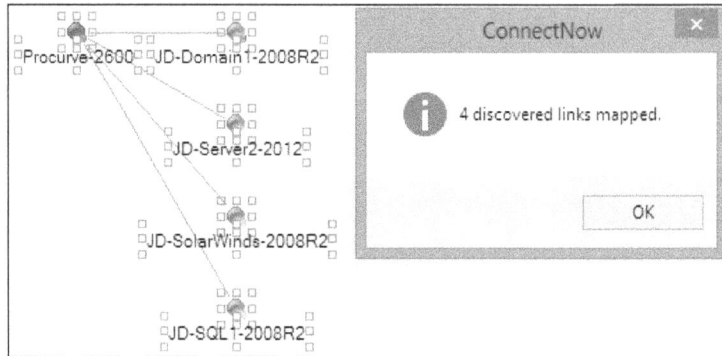

Network Atlas will automatically link the objects using a straight line and inform you how many objects it was able to link together. **ConnectNow** only links objects together and it does not arrange the objects in the edit pane for you. It is up to you to arrange the linked nodes in a manner you deem appropriate. The good news is that, once an object is linked via a line, you can use the Select tool to move the object around the map without having to worry about reconnecting the line. The line will permanently remain connected to the objects.

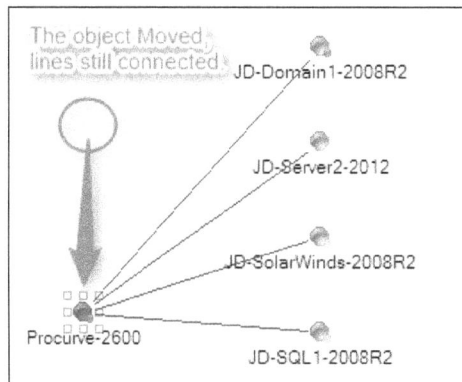

After you are done making your changes to the map, click on the **Save** button on the top-right corner of the window (indicated by a floppy disk icon). Choose a name for the newly created map and click on **OK**.

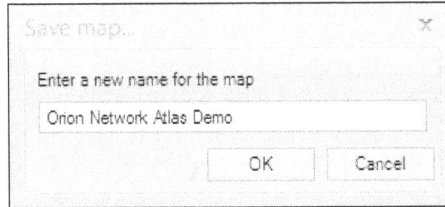

The new map will be listed in the **Maps** view within the objects pane.

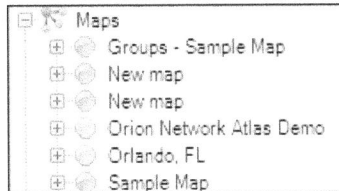

Applying maps to Orion dashboard views

So, now that the new map has been created, it is time to apply the map to a view in the Orion dashboard. There are many different views in Orion NPM and all of them can be customized to your liking. In the following example, we will see how to apply a basic network map to the **Home Summary** view in Orion NPM.

Log into the dashboard and click on the **Home Summary** tab. Under the **Maps** module, click on the **Edit** button. Notice that your new map appears in the list! Click on the map title to highlight it, make any other necessary changes to the resource, then click on **SUBMIT**.

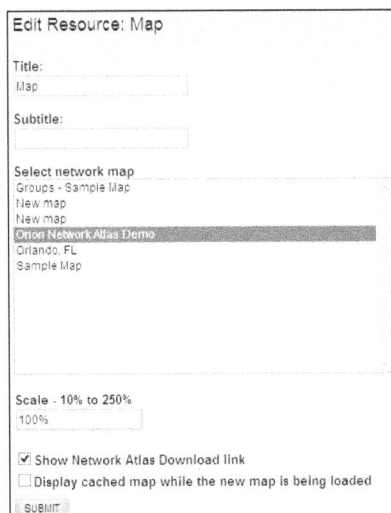

> If your map is very large or has a great deal of objects in it, it may be a good idea to check the box for **Display cached map while the new map is being loaded**. This will force the web browser to load the Orion dashboard faster by using the local cache for the map instead of waiting for the map to update first.

Your new map will now appear in the **Home Summary** view and will be one of the first resources you see when you log into Orion NPM.

Adding network maps to Orion dashboard views not only makes it look more professional, but also can provide some quick insight to issues on the network. This is only one example of how to build a basic network map and how to apply those custom maps to dashboard views in Orion NPM. The Orion Network Atlas is capable of displaying thousands of different objects with different relationships. And you can build different network maps for different purposes. The sky is the limit! You will definitely want to put some time into building a custom map for your own environment or customers.

Summary

So, we have now reached the end of the chapter. We discussed the various group, node, and NPM reports, how to build reports using Report Writer, and how to schedule automatic report delivery with Report Scheduler. We also discussed the network mapping utility, Orion Network Atlas, and how to apply those maps to views in the Orion Dashboard.

In the next chapter, we are going to discuss some of the things that are critical to the operation and maintenance of Orion NPM. We are going to discuss backups, upgrades, migrations, licensing, database cleanup, MIB databases, and performance and optimization. So let's turn the page and get started with the discussion, shall we?

8
Maintenance

My family and I live in Florida. Living in Florida has its perks, but it also has its downsides. For example, I must mow my lawn every week or two in order to keep it from looking like my house is in the middle of the jungle. When I don't mow my lawn as often as I should, it takes me twice as long to complete the job, and mowing extremely long grass wreaks havoc on my lawnmower's blades.

Not only do I need to mow my lawn, but I need to perform some basic maintenance on the mower itself. I need to check the fuel levels, change the oil from time to time, check the spark plug, and more. Over time, I stopped mowing my lawn as frequently as I needed to, I stopped performing maintenance on the mower by not changing the oil or checking the spark plug, and I basically let everything slide. One day, I finally decided to get outside and mow the lawn but the motor refused to turn over! I couldn't cut the already very long grass. Long story short, I had to pay to get the mower fixed because I didn't perform regular maintenance.

Just about everything in life needs maintenance, including your Orion NPM system. SolarWinds provides several tools to help maintain the Orion NPM system and keep it running at its peak. Not all forms of maintenance are the same. For example, one of the maintenance tools in Orion NPM is a database cleanup utility which removes invalid items from the database and compacts it to save storage space. Another form of maintenance is migrating the Orion NPM installation to a new Windows Server. And yet another form of maintenance is enabling and disabling services. All of these tasks assist with maintaining your Orion NPM server. This chapter discusses the various maintenance tools built into an Orion NPM, best practices of maintenance, and the Orion NPM migration processes.

Service management

SolarWinds Orion NPM includes a very simple tool to assist with stopping and starting the various Orion NPM services running on the Windows Server where Orion NPM is installed on. This utility is called the **Orion Service Manager**. You can find this tool in the Start menu under **Advanced Features** in the **SolarWinds Orion** folder.

The utility itself is very straightforward and easy to understand. The right-hand side of the window displays if the service is running or if it has been stopped. The left-hand side of the window displays the service's description. Depending on your network environment and needs, it may be normal to see one or two of these services completely stopped if they are not in use (such as the **Syslog** service).

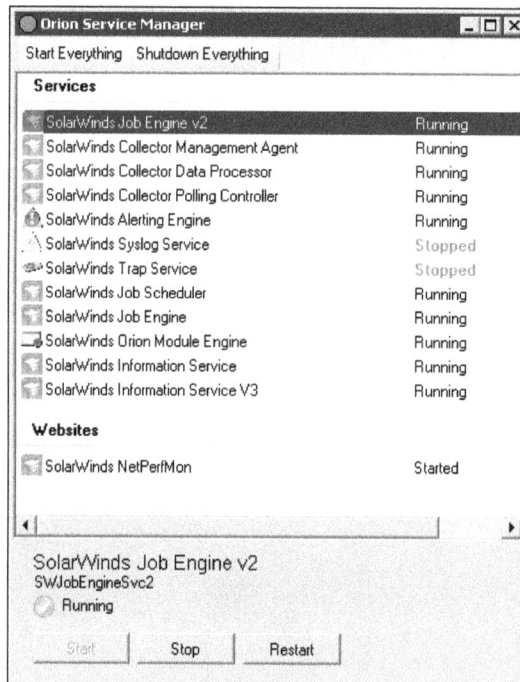

There are two buttons at the top of the window that will start or stop all of the Orion NPM services. To start or stop a single service, simply click on it to highlight it and click on the **Start**, **Stop**, or **Restart** button at the bottom of the window.

One of the least documented things about this utility is that it actually shows the real name for the Windows Service at the bottom of the screen! Take a look at the following example:

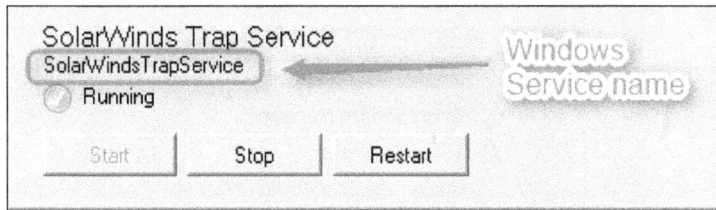

The Windows Service name for the SolarWinds Trap Service is **SolarWindsTrapService**. Knowing this, I can use the command line to start or stop the service using the NET utility. For example, if the Trap Service is stopped, I can open an elevated command line and execute the following:

```
net start SolarWindsTrapService
```

Armed with this information, you can create custom batch files or scripts to perform some common tasks against SolarWinds Orion NPM services.

Database maintenance

SolarWinds Orion NPM is a SQL database-driven application. It stores all information about every monitored node, all historical information, alert settings, user settings, and every other piece of information that Orion NPM needs to remember in the SQL database. To have Orion NPM perform at its peak, you need to make sure that the database is optimized, regularly backed up, and free of inconsistencies. Two database tools are included in each Orion NPM installation. These are the **Database Manager** and the **Database Cleanup** utilities.

Database Manager

The Database Manager application is used to perform several different types of tasks, from manually editing tables in the database, performing a database backup, restoring a backup, compacting a database, and creating a backup schedule. This is the tool that you will use to perform most database maintenance tasks.

To launch the Database Manager, log into the Windows Server that Orion NPM is installed on and navigate to **Start | All Programs | SolarWinds Orion | Advanced Features | Database Manager**.

If this is the first time you have launched the Database Manager application, you will need to add your SQL Server. Click on the **Add Server** button in the **Database Manager** toolbar. Enter the SQL Server information in the textbox and choose the login method and click on the **Connect to Database Server** button to continue.

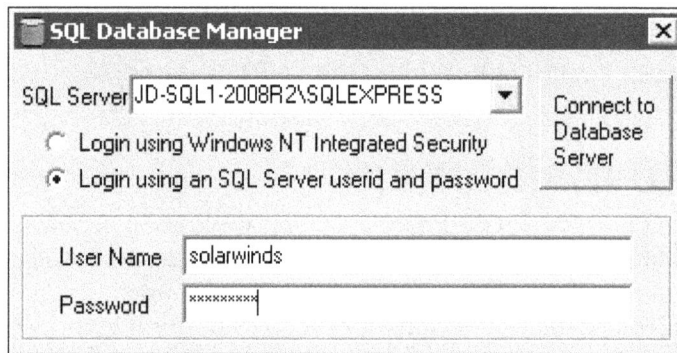

Once you are connected to a SQL Server database, it will be displayed in the left-hand view pane within **Database Manager**.

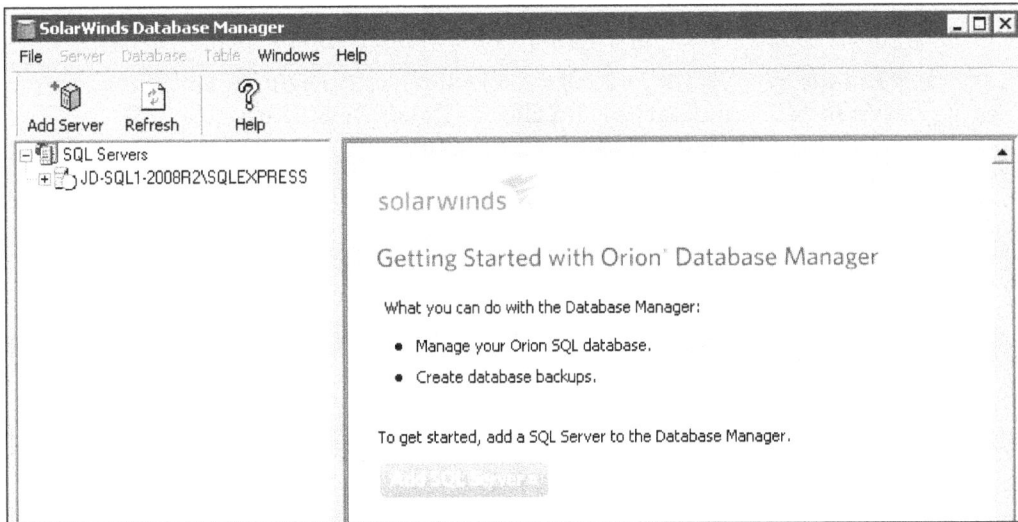

Now that the SQL Server has been added, we can perform several tasks. I will show you how to perform several routine tasks including backing up the database, setting up a database backup schedule, verifying a database backup file, and restoring the database.

Creating a database backup schedule

SQL databases can store a great deal of data and they are very helpful when needing to process large amounts of data. But databases cannot read minds—it is up to you to create a backup schedule for a database and define where the backups will be stored.

Creating a database backup schedule should be one of your first tasks after configuring your Orion NPM installation. The last thing you should worry about is losing your database in the event of a catastrophic failure. To create a backup schedule for your database, follow the given process:

1. Expand the SQL Server in the left-hand side pane. Right-click on the **SolarWindsOrion** database and choose **Database Backup Schedule**.

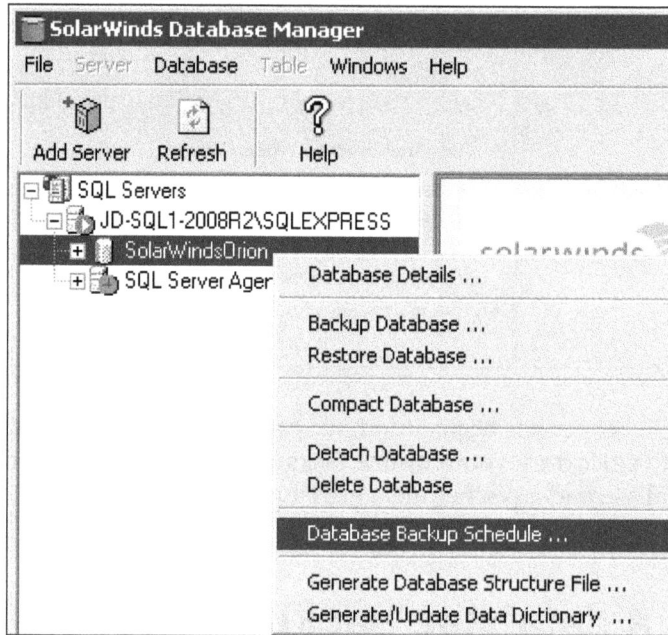

2. The **Backup Schedule** wizard appears. Choose an appropriate time schedule and click on **Next** to continue.

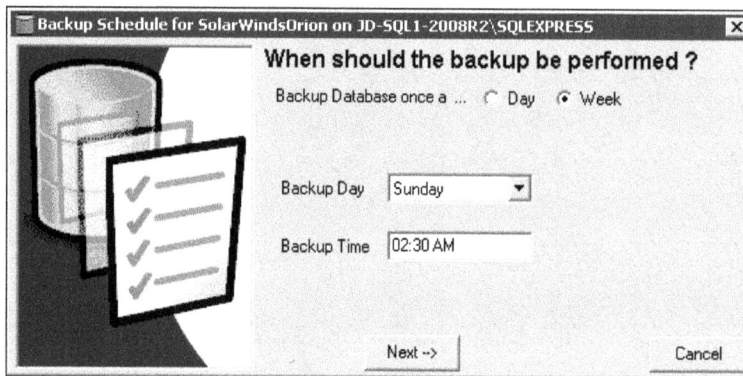

Backing up the database once per week should be more than sufficient for almost every Orion NPM deployment. However, for networks that have frequent or more dynamic changes, it may be appropriate to configure the backup to run more frequently. Just make sure that you configure the backup day and time to be during your organization's maintenance hours, since running a backup can take a few minutes and may impact the performance of both Orion NPM and your SQL Server.

3. Choose whether or not to compact and shrink the database before backing up the database. This option performs the same process as that of the database cleanup utility. It is recommended to leave this option enabled as the utility will clean up the database first before saving a backup. Click on **Next** to continue.

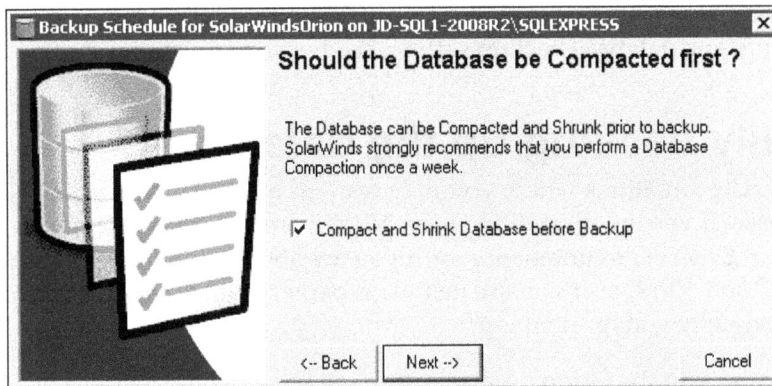

4. The final screen in the **Backup Schedule** wizard is to choose where to save the backup files and whether you want to save a report of the backup job for verification purposes. After making your selections, click on **Finish** to save the job.

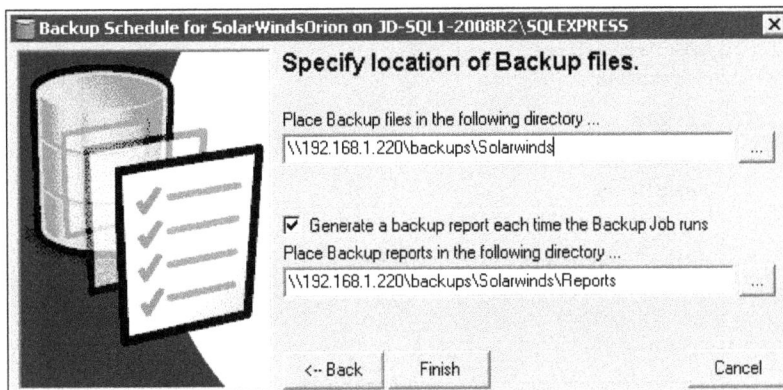

5. Once the backup job has been created, a notification window will appear as shown in the following screenshot:

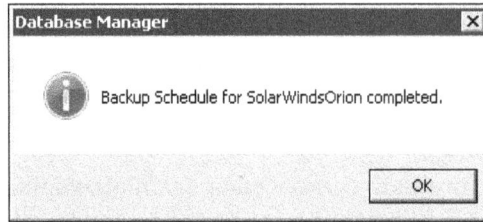

You can always go back and edit the backup schedule by performing the same steps in this section. One thing to remember about database backup schedules in Orion NPM is that you cannot create multiple backup schedule operations on the same Orion NPM server even if there are multiple Orion NPM databases. You can configure only one database backup setting, so choose your backup options wisely.

Manually backing up the database

There are certain situations where you may want to manually backup the Orion NPM database. If you are migrating Orion NPM from one server to another, performing SQL Server maintenance, or if you are about to make a great deal of changes to Orion NPM, or if you are just plain paranoid and want a backup, you can do so manually and on-demand.

To manually backup the Orion database, right-click on the database in the **Database Manager** utility and choose **Backup Database**. Verify that the correct database is selected in the first drop-down box, type a description for the backup file, and choose a location and filename for the backup. When creating a backup file, you can choose to append to the end of an existing backup file or overwrite an existing backup file. Once you have made your selection, click on **OK** to create the backup file.

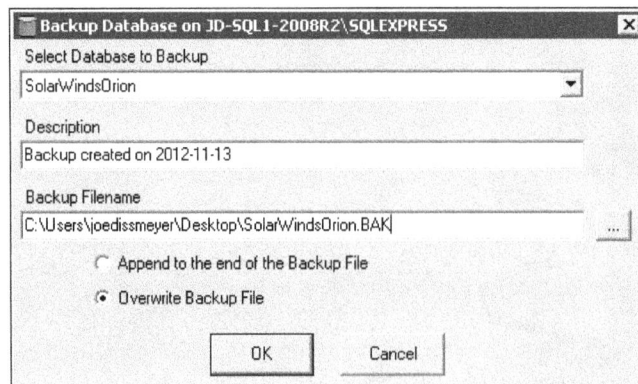

Once the database backup has completed, a notification will appear informing you if the backup was successful or not. If it was not successful, you may have chosen an incorrect database. Attempt to create a new backup and make sure that the correct database is selected.

> If you choose to save a backup file to a local drive (C:\, D:\, and so on), database backup files are created on the SQL Server itself—*not* the Windows Server that Orion NPM is installed on. This means, if you choose to save the backup file to C:\, Database Manager will save the file to C:\ on the SQL Server where the database resides. The reason why this happens is due to the application's design and is a limitation of the database backup utility. It is recommended to store database backup files on a network file server.

Restoring the database

There may be a time where you need to restore the Orion NPM database because a large number of nodes were accidentally removed from the database, the Orion NPM database is corrupt, or you are moving the database from one server to another. Some other reasons may be the need to recover from a catastrophic failure of either Orion NPM or Microsoft SQL Server, and so on. Thankfully, restoring the Orion NPM database is a relatively easy task to perform.

There are a few things to keep in mind when restoring an Orion NPM database. First, stop all Orion NPM services before starting a database restore. Next, if a restore is going to overwrite an existing Orion NPM database, the SQL Server service needs to be restarted on the Windows Server where Microsoft SQL Server is installed. Doing this will forcefully terminate all connections to the database from other sources (if there are any) and will ensure that the restore succeeds on the first attempt.

> Restarting the SQL Server service will impact connections to other databases as well, so it is recommended to perform database restoration tasks during maintenance hours.

Perform the following steps to restore a backed up database:

1. Launch the **Database Manager** application.

2. Right-click on the Microsoft SQL Server instance and choose **Restore Database**.

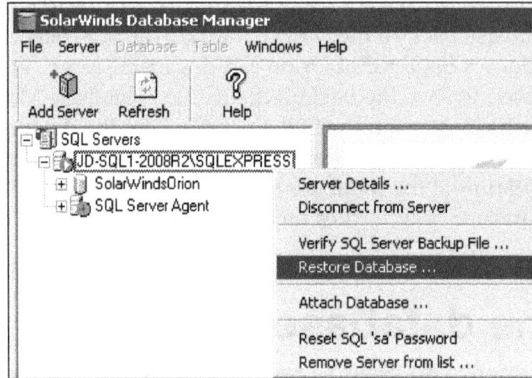

3. The database restore wizard will appear on screen. Enter the location of the database backup file then click on the **Verify** button.

4. The utility performs a quick check on the backup file to make sure it is not corrupt. If it passes the verification, a notification window will appear and all data will automatically populate in the wizard screen as shown in the following screenshot. If the verification does not pass, choose a different backup file and try again.

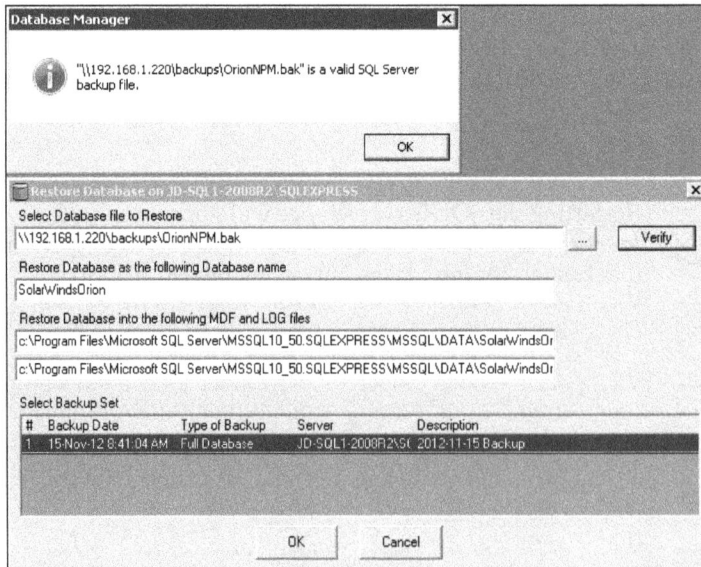

If you have additional database backups to restore (such as an incremental backup file), enter the location of the file and click on **Verify** again. The additional backup file will appear in the **Select Backup Set** display grid. Click on **OK** to continue.

5. Depending on the size of the database, restoration can take several minutes to complete. Click on **OK** to finish the restore.

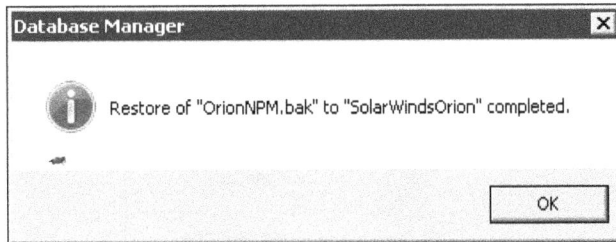

6. Launch the **Orion Service Manager** and start all Orion services.

7. Open the Orion dashboard and verify everything is in order. If the dashboard does not display correctly, you may need to restart the server to resolve the issue.

If you receive an error similar to the following screenshot, you didn't restart the SQL Server database service before attempting to perform a restore. Restart the SQL Server service on the Microsoft SQL Server and try again.

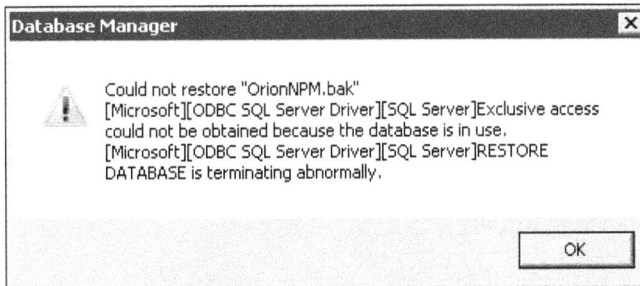

Database clean-up

Orion NPM includes a utility that will automatically compact and remove invalid database entries called **Database Maintenance**. To launch the **Database Maintenance** utility, log into the Windows Server that Orion NPM is installed on and navigate to **Start | All Programs | SolarWinds Orion | Database Maintenance**.

To start the database maintenance process, click on the **Start** button. The application will run through the cleanup process and inform you when it is completed.

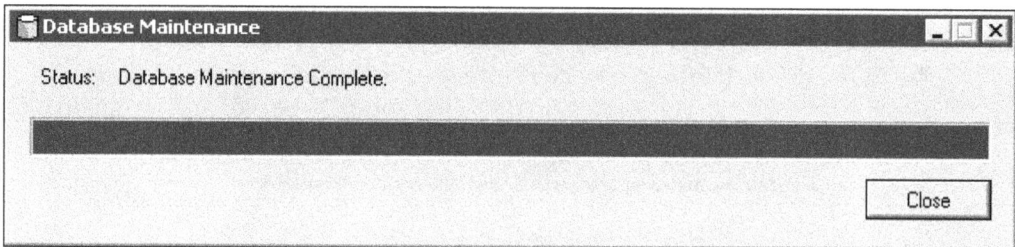

It is a good idea to run this utility at least once per month. Unfortunately, there is no scheduling tool that will automatically run this tool on your behalf. You must run database maintenance manually.

License management

It is important to keep on top of your available licensing in Orion NPM in order to assist with capacity planning and budgeting. If your organization is adding more devices to the network, and you need to monitor the resources within these new devices, it is critical that you know how much licensing is in use on your Orion NPM system and plan on adding more licensing to it. There are a few different ways to see if you need to order more licensing in Orion NPM. The first is by viewing the **License Details** view in Orion Web Administration.

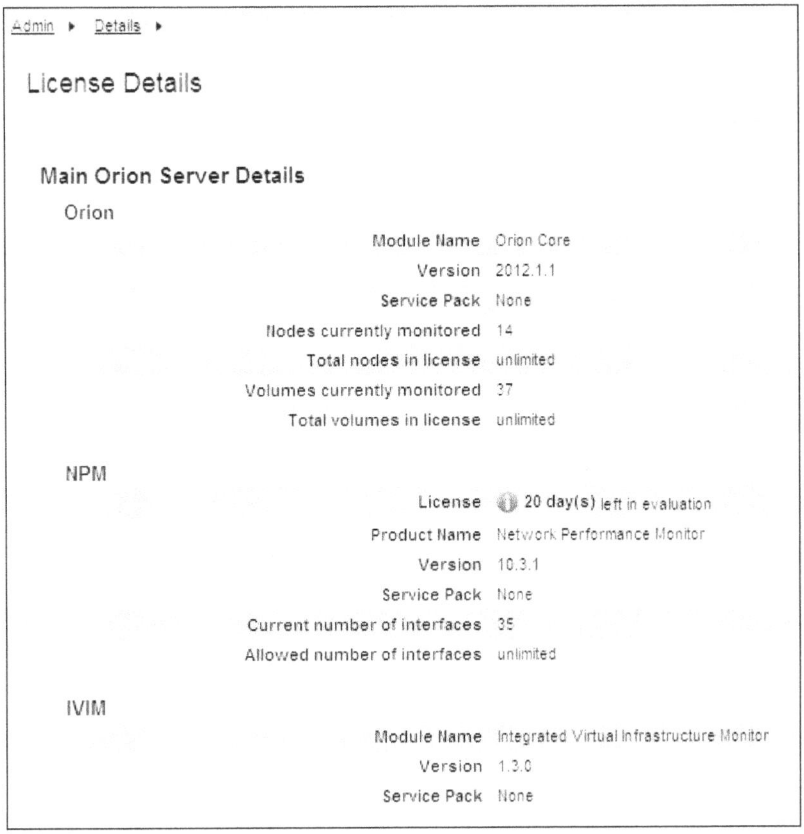

Plenty of details about Orion NPM licensing are displayed here. There are the following three primary sections in this view:

- **Orion**
- **NPM**
- **IVIM**

The **Orion** section is focused on detailing the licensing for the node and volume elements. It displays the version of Orion Core that is installed, the number of nodes currently being monitored by Orion, total amount of nodes in the license, the number of volumes being monitored, and the number of volumes allowed in the license.

The **NPM** section focuses on the interface licensing as well as the actual licensing state of the Orion NPM install itself. **License** displays the installed license of the server. If my installed license is the SL100 pack, then "SL100" would be displayed in this field. Also displayed is the current number of monitored interfaces and the number of licensed interfaces.

The final section in this view is **IVIM**, which stands for **Integrated Virtual Infrastructure Monitor**. **IVIM** is actually a small subset of the Orion NPM add-on **Orion Virtualization Manager** and there is no special licensing needed in order to use it. **IVIM** is the module that handles the monitoring of VMware and UCS nodes.

Another way to view the licensing of Orion NPM is by using the **License Status** tool, which is an application installed on the Orion NPM Windows Server. This tool only displays the actual licensing state of Orion NPM, not the number of licenses in use at this moment in time. You can find it by navigating to **Start | SolarWinds Orion | Network Performance Monitor | Network Performance Monitor Licensing**. In it, you can view the activation key used to license the server, the license type (SL100, SL500, and so on), and the number of days left in your evaluation (if you are running an evaluation installation).

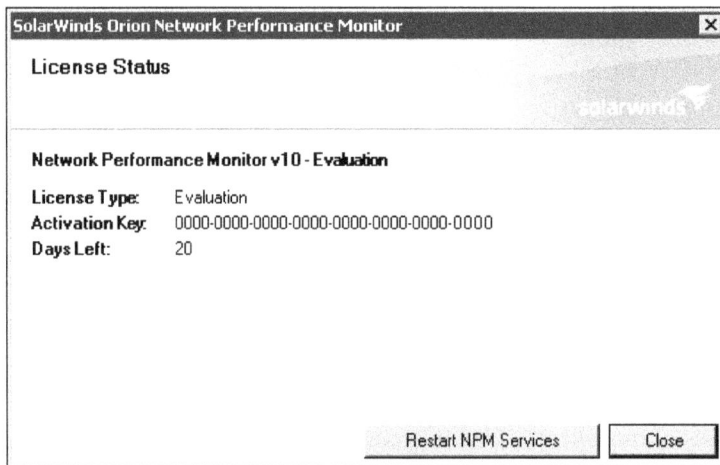

Other than viewing the license key and basic license information, you can restart the Orion NPM services from here as well.

Using the SolarWinds License Manager tool

The **SolarWinds License Manager** is a tool provided to you for self-service licensing activation after SolarWinds has been installed and configured for the first time. This is the tool that you use to change or activate licenses. It is also the tool you use to deactivate a SolarWinds license. There are many different reasons for needing to deactivate a license, such as decommissioning a SolarWinds Orion server, moving an Orion NPM installation to a new server, or re-assigning a license key to a different server within the organization. Whatever the need is for, SolarWinds provides this tool so you can activate and deactivate licenses without having to contact SolarWinds support.

The **License Manager** application is not installed by default. After logging into the Windows Server that Orion NPM is installed on, click on **Start** | **All Programs** | **SolarWinds** | **SolarWinds License Manager Setup**.

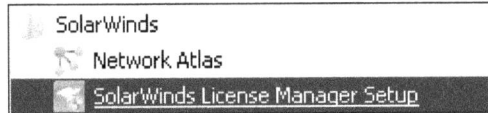

Accept the license agreement, click on **Install**, and allow the application to install on its own. Once installed, you can launch the application from **Start** | **All Programs** | **SolarWinds** | **SolarWinds License Manager**.

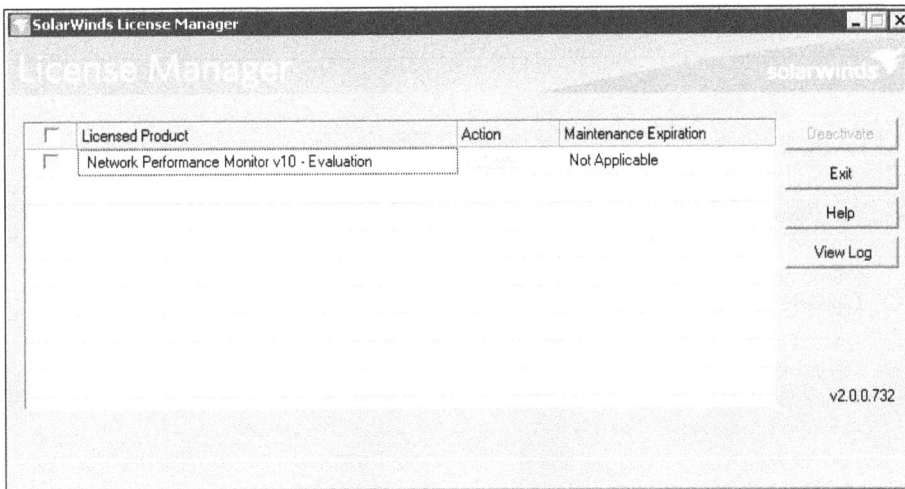

License Manager displays all licensed products within the display window. As described in *Chapter 1, Installation*, when you install a SolarWinds product, it is automatically placed in 30-day trial mode. To activate a product, click on the **Activate** link next to the product's description in **License Manager**. Simply follow the activation wizard from this point to activate the product. To deactivate a license, place a check mark next to the product you wish to deactivate and click on the **Deactivate** button.

Polling engines

Each Orion NPM installation is itself a singular polling engine and is rated to be able to poll up to 10,000 elements without impacting performance. However, there may be a day where your polling engine approaches, or exceeds, its polling rate limit. What do you do when this happens?

First you should take a look to see if the current polling engine is overloaded. To do this, open the Orion Web Administration page and click on **Polling Engines** in the **Details** section.

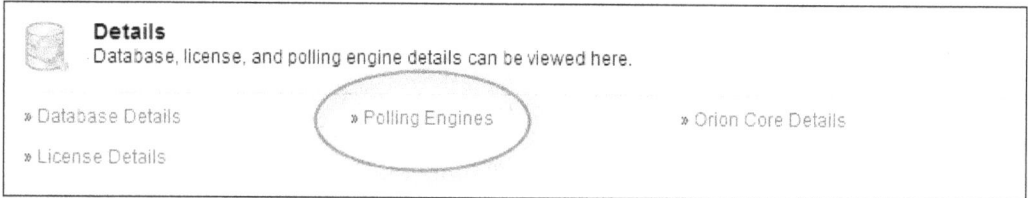

This will display the **Polling Engines** view that shows all of the details about a particular polling engine.

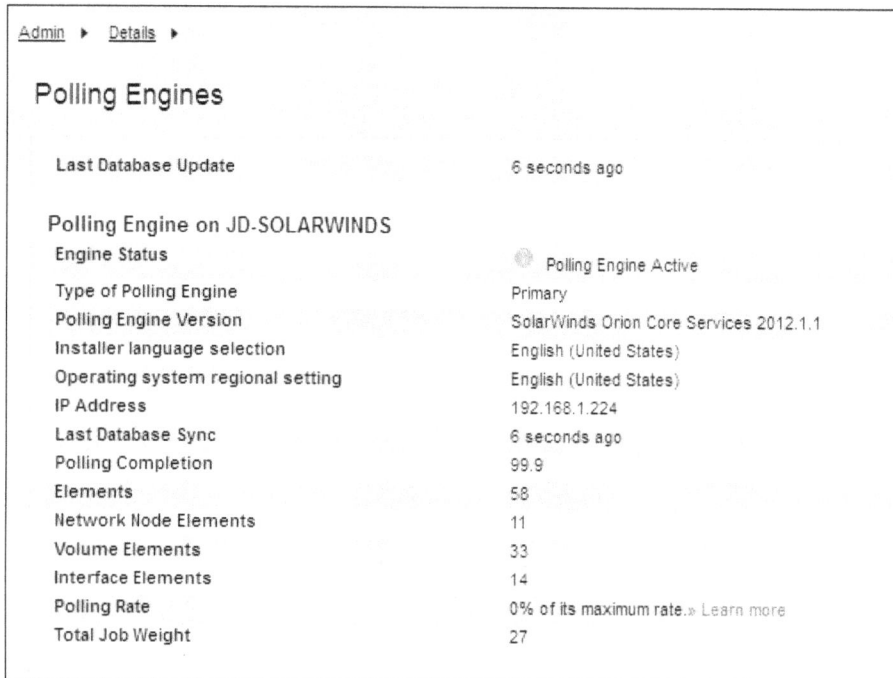

What is important to note from this view is the polling rate. When the polling rate hits or exceeds 85 percent, Orion NPM will display a warning in the top menu bar that says **Poller Status Warning: A poller is either approaching or in excess of its polling rate limit**. Orion NPM attempts to compensate for this by increasing the polling time for each node automatically. Polling will complete, just at a later time than expected. In order to fix this problem there are two things you can do.

The first is to manually adjust the polling rate. If you recall, you can define the polling rate for each node, interface, and volume from each node's properties.

```
Polling

              Node Status Polling:  120                    seconds
             Collect Statistics Every:  10                     minutes
        Poll for Topology Data Every:  30                     minutes
                  Polling Engine:  No polling engine selected
```

The default settings for polling are 120 minutes for status polling, 10 minutes for collecting statistics, and 30 minutes for topology data. You can keep the polling time low for critical nodes, but increase the polling time on lesser critical nodes. If you have adjusted the polling settings for any of your nodes to a lower value, this can severely impact not only the polling performance but possibly your network as well. Increasing the polling settings to higher values will free up polling resources and reduce the polling rate.

If adjusting the polling rates for your monitored nodes does not resolve the issue, or if you are unable to do so due to contracts or agreed upon service-level agreements, the second option is to add an additional polling engine to the network. What this really means is that you install another licensed Orion NPM instance on a second Windows Server and have it linked to the same SQL Server database your primary polling engine is connected to. You can have as many Orion NPM installations deployed throughout your organization as you need and tie them all together by sharing the same SQL Server database. Then, you can modify a node's properties to define which polling engine it will report to.

The following are the things you should know about additional polling engines:

- Only one polling engine is installed per Orion NPM server. There is no way to increase this amount per server.
- Every time you upgrade or update the primary polling engine, you must also upgrade all secondary polling engines since they are connected to the same SQL database.
- The primary polling engine need not be available in order for a secondary polling engine to collect data and save it to the shared SQL Server database.
- Alerts are controlled from the primary polling server.
- The primary polling server does not need to be offline in order to install a secondary poller.

MIB database updates

Technology changes faster than we can all keep up with sometimes. New smartphones, computers, tablets, network devices, and software are released almost every day which makes it difficult to keep up with changes. In the network and server world, new equipment with new software is also released every day. With these new devices comes new information (or OIDs) for monitoring them using Orion NPM. Orion NPM does a pretty good job of being able to monitor just about every single network device using SNMP, since every device has some type of standard OIDs that can be polled. But there are still some aspects of a device that you may want to monitor that Orion NPM cannot poll for out of the box. This is where you will need to update the Orion MIB database.

By default, the standard SolarWinds Orion NPM installation does not come with its full MIB database. This is because a fully-compiled MIB database for Orion is now approaching 700 MB! Instead, it is up to you to add this complete database to your own Orion NPM installation. SolarWinds releases a new compiled MIB database every month and makes it available on their website for free at `http://downloads.solarwinds.com/solarwinds/Release/MIB-Database/MIBs.zip`.

Alternatively, you can log into the SolarWinds Customer Portal using your customer account and click on the **Orion MIB Database** download link under the **Helpful Links** section.

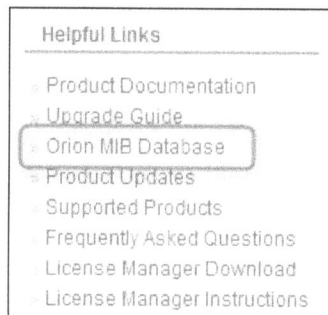

```
Helpful Links

  Product Documentation
  Upgrade Guide
  Orion MIB Database
  Product Updates
  Supported Products
  Frequently Asked Questions
  License Manager Download
  License Manager Instructions
```

The following are some things to know about MIB databases and SolarWinds Orion NPM:

- You cannot download the MIBs from a vendor and manually add them to Orion NPM. SolarWinds Orion is locked down to allow only the master MIB database compiled by SolarWinds support staff.

- If you have a device where its MIBs are not included in the master SolarWinds Orion MIB database for Orion, contact SolarWinds support to have them include your device in the next MIB database compilation.

Once you have the ZIP file, extract it and copy the contents of it to a temporary file. Inside the ZIP file, there are the following two files:

- `MIBs.cfg`
- `Readme.rtf`

The readme file is a rich-text document containing instructions on how to apply the new MIB database to your Orion NPM installation. The `MIBs.cfg` file is the brand new database. Perform the following steps to apply the new MIB database to your Orion NPM installation:

1. Stop all Orion NPM services using the **Orion Service Manager**.
2. Copy the new `MIBs.cfg` file to the SolarWinds installation folder on the Windows Server.
   ○ If Orion NPM is installed on Windows Server 2003, copy the `MIB.cfg` file to `C:\Program Files\SolarWinds\Common\`.
   ○ If Orion NPM is installed on Windows Server 2008 and above, copy the `MIBs.cfg` file to `C:\ProgramData\SolarWinds\`.
3. Start all Orion NPM services using **Orion Service Manager**.

The new MIB database will automatically be applied to your installation. Your job is now complete! Now that the new database has been applied, you can open up the Universal Device Poller application and create custom pollers for your device.

Migrating Orion NPM

Moving database servers and software installations back and forth between old servers and new servers is a common practice in the enterprise. In order to retire old and outdated server hardware, various software and operating systems need to be migrated to newer hardware or migrated to a virtualized server environment. I promise you that there will be a day where you will need to perform some type of migration for your Orion NPM server. There are the following two different types of migration processes for Orion NPM:

- Orion SQL database migration
- Orion NPM installation migration

The migration process involves ramping up a second Orion NPM installation as a second polling engine, linking the existing SQL database to the new install, configuring each node to use the new polling engine, then moving the licensing from the old install to the new one. The following is a migration diagram:

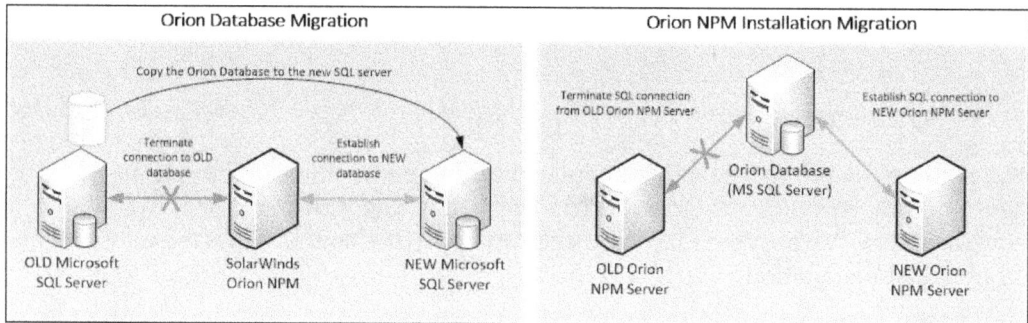

It is important to know that an Orion NPM migration does not move or reconfigure the Orion SQL database. Also, the Microsoft SQL Server housing the Orion database remains in place as is and a new Orion NPM installation is ramped up on a new, or different, Windows Server.

> When migrating an Orion database or Orion NPM installation, the version of Orion NPM that you install on the new server must be identical to the version on the old server. This is due to database incompatibilities between software versions. If you are planning on installing a newer version of Orion NPM on the new server, upgrade the old installation first.

Moving the Orion database

The process for moving the Orion database involves backing up the existing database, restoring it on the new SQL Server (or existing SQL Server), running the Orion Configuration Wizard, and re-linking the new database to the existing Orion NPM installation.

> Before moving the Orion database to a new SQL Server, make sure that your new SQL Server is properly configured and meets the minimum requirements. If you need to, go back to *Chapter 1, Installation*, and review the SQL Server software and hardware requirements.

Duplicating the Orion database

The following is a detailed procedure for moving an Orion database from one Microsoft SQL Server to a new Microsoft SQL Server (or a different one).

1. Log into the Windows Server where Orion NPM is installed and using the **Orion Service Manager**, stop all Orion NPM services.

2. Close the **Orion Service Manager** application then launch **Database Manager**.

3. Click on the **Add Server** button in the toolbar and add both the Microsoft SQL Server instances to the SQL Server list in the left-hand side view pane. Notice that the new server does not have a copy of the Orion NPM database.

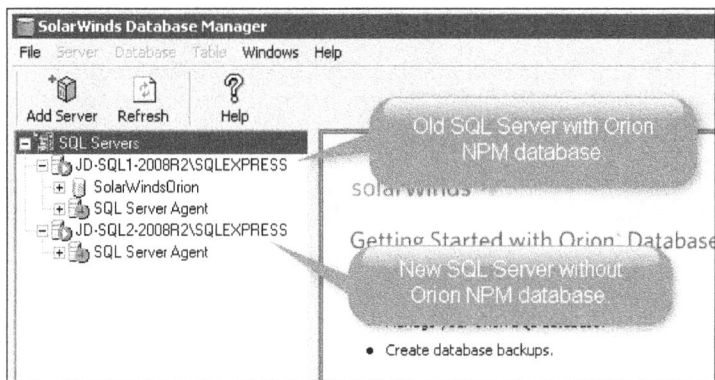

4. Right-click on the old server and choose **Backup Database**.

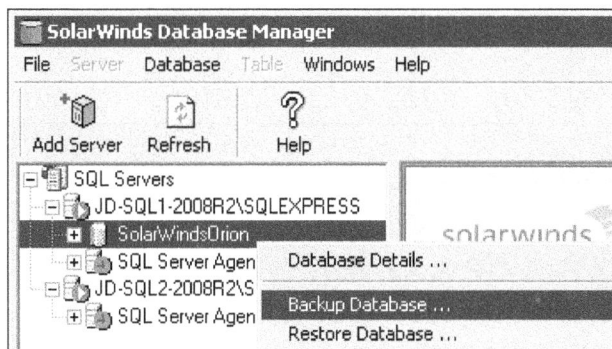

5. Select a location to save the file then click on **OK** to create the backup file.

6. Right-click on the new SQL Server instance and choose **Restore Database**.

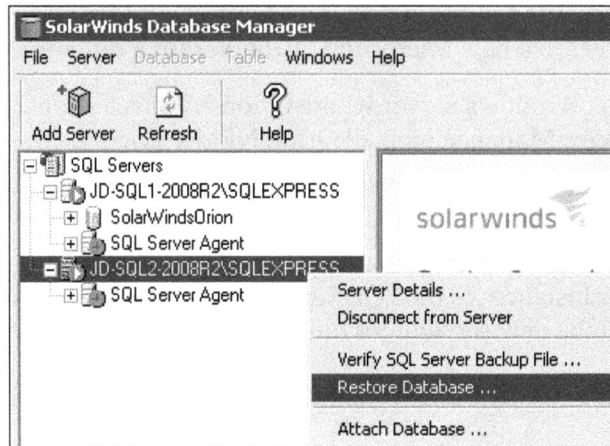

7. Enter the location of the backup file then click on the **Verify** button.

8. After the verification is successful, click on **OK** to restore the backup to the new SQL Server database.

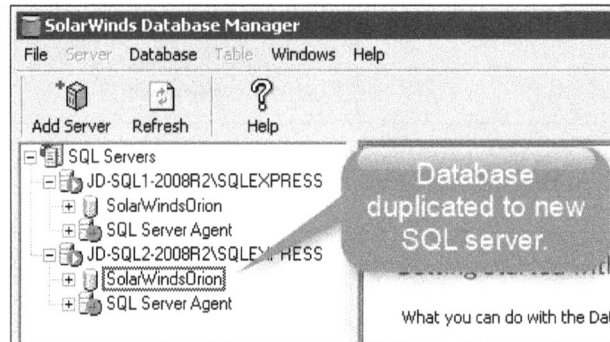

9. Close the **Database Manager** application.

Configuring the new Orion NPM server

Now that the database has been duplicated to the new SQL Server, you need to run the **Orion Configuration Wizard** and point Orion NPM to the new SQL database.

1. Launch the **Orion Configuration Wizard**.

2. Place a check mark next to the **Database** option then click on **Next**.

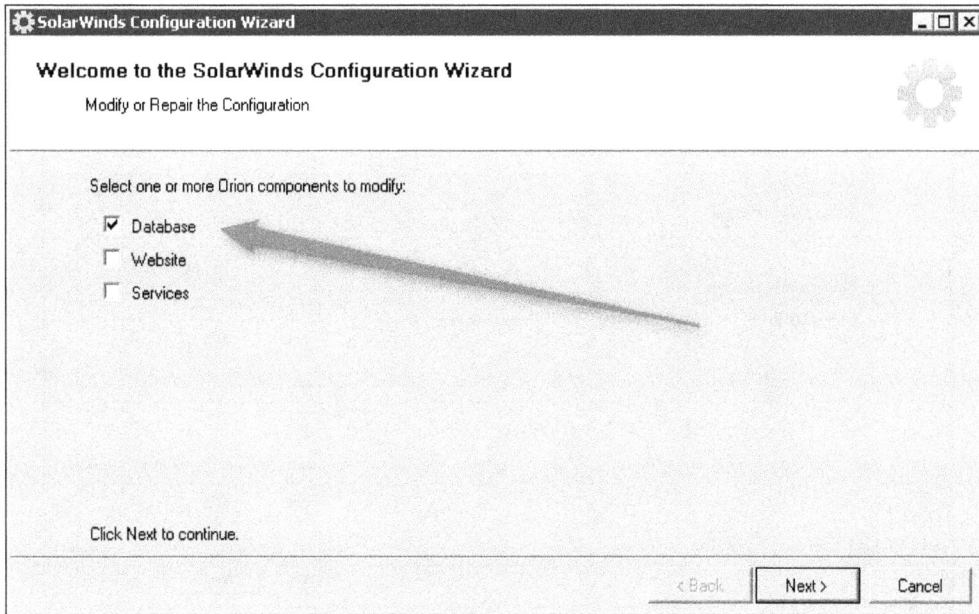

3. Define the new SQL Server instance, enter the local SQL Server credentials, then click on **Next** to continue.

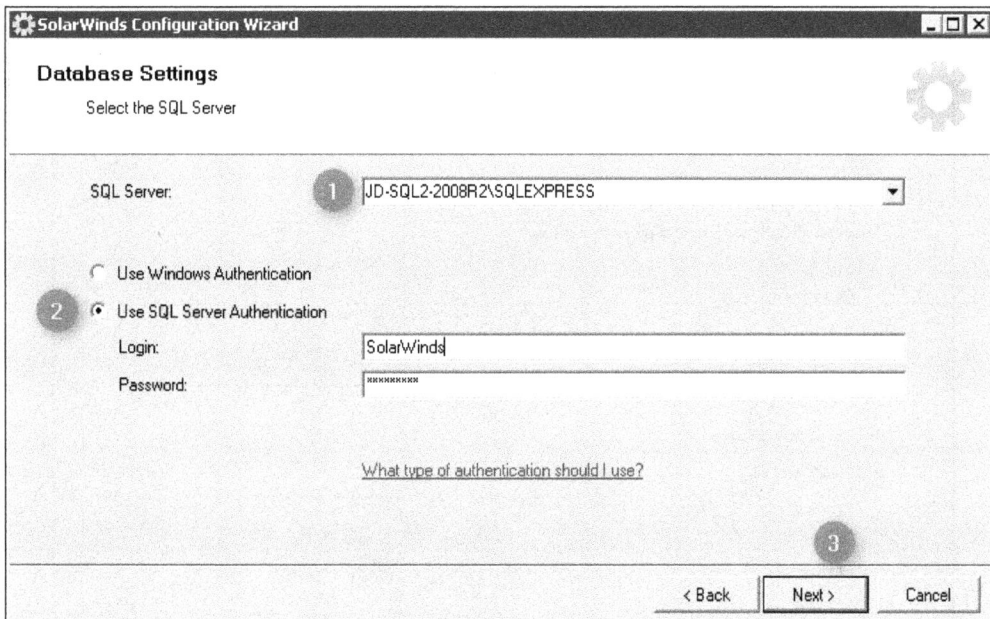

4. Select the existing SolarWinds database then click on **Next**. The database name will be identical to the database name from the old SQL Server.

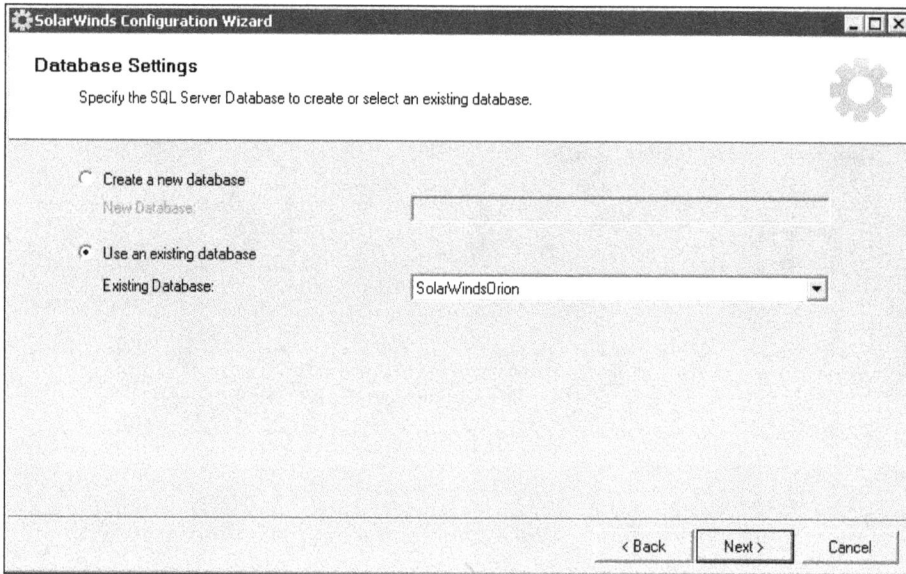

5. At the **Database Account** screen, choose **Use an existing account** and select the local SQL user account for SolarWinds. Click on **Next** to continue, then again on **Next** one more time to configure Orion NPM for the new SQL database.

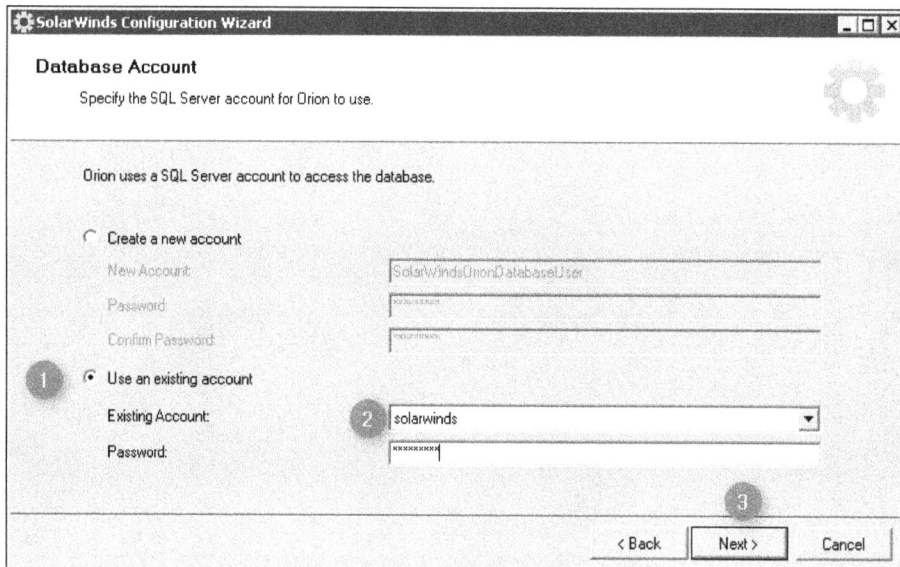

6. The configuration wizard will take a few minutes to point the Orion NPM installation to the new SQL Server. When the job is complete, click on **Finish** to close the wizard.

7. Restart the Orion NPM server to complete the transition.

This concludes the process of moving an Orion database to a new Microsoft SQL Server instance. The next section discusses how to move an Orion NPM installation to a new server.

Moving an Orion NPM installation

As described before, you can move an Orion NPM installation from one server to another. Provided the existing Orion NPM database is accessible, you can move forward with setting up a new Orion NPM installation on your new server. The great thing about Orion NPM is that just about everything is embedded in the Orion database. The only items that are not included in the Orion NPM database are the product activation and custom reports. Both need to be migrated manually. We will discuss this throughout this section.

Configuring the new Orion NPM installation

The process for setting up your new Orion NPM installation is almost identical to what is described in *Chapter 1, Installation,* with the exception of choosing an existing SQL database in the configuration wizard.

1. Perform a brand new installation of SolarWinds Orion NPM on a Windows Server and launch the **Orion Configuration Manager**. Click on **Next** to get started.

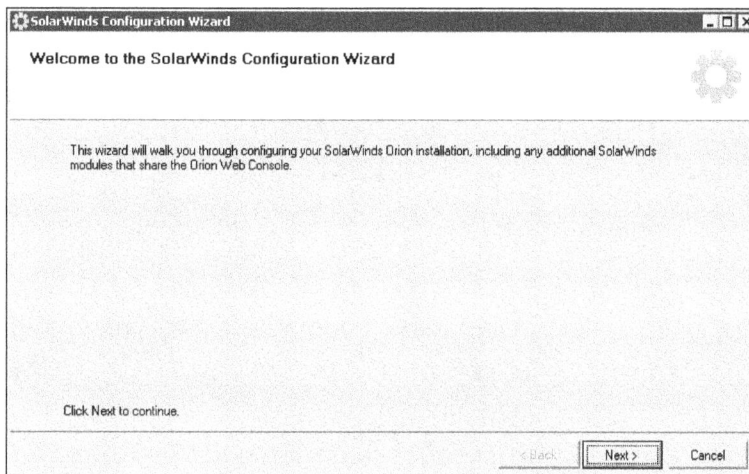

2. On the **Database Settings** screen, choose the SQL Server instance and local user account. Click on **Next** to continue.

3. Choose the existing Orion NPM database on the SQL Server and click on **Next** to continue.

4. A warning window will appear. This warning informs you that another polling engine (which is the other Orion NPM installation) was detected as being active on this database. After this new Orion NPM installation is properly configured, we will be removing the old Orion NPM poller from the database. Click on **Yes** to continue.

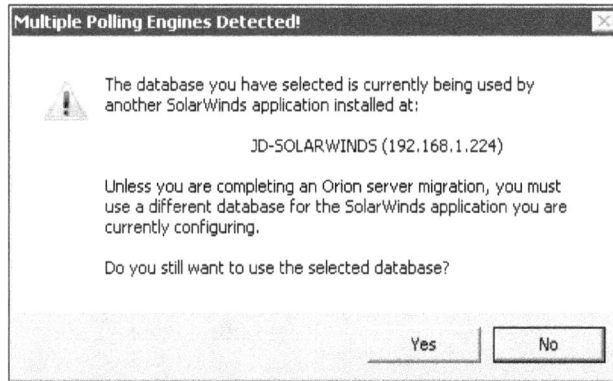

5. Another warning window will appear re-iterating the same information from the previous warning. Click on **OK** to continue.

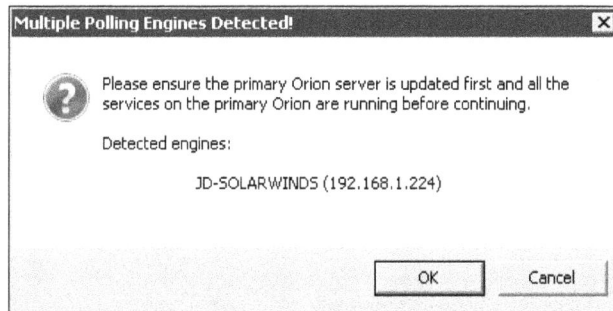

Details on the remaining settings in the **Orion Configuration Wizard** are covered in the first two chapters of this book.

6. At the **Database Account** screen, choose **Use an existing account**, define the local SQL Server account for SolarWinds, then click on **Next** to continue.

7. Define your website settings and click on **Next** to continue.

8. Choose the services to install then click on **Next** to continue, and click on **Next** once more to finish the configuration.

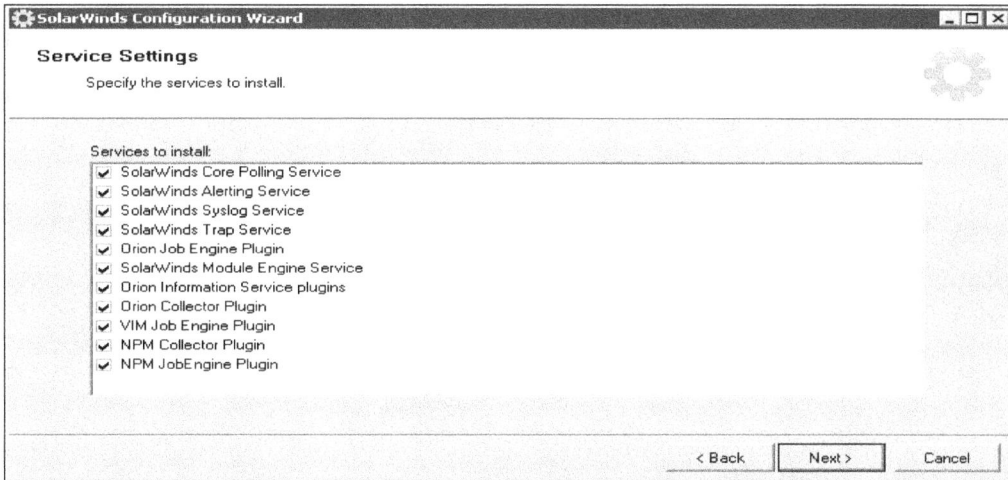

9. After a few minutes, the new Orion NPM installation will be up and running. Close the **Configuration Wizard**.

Re-assigning polling engines

The new Orion NPM installation is now up and running. Next, you need to open the **Database Manager** application and execute a command to re-assign all nodes to the new polling engine.

1. Stop all Orion services using the **Orion Service Manager** from the Start menu.

2. Launch **Database Manager** from the Start menu. Add the Microsoft SQL Server to the list.

3. Expand the SQL Server instance name on the left-hand side view pane then expand the Orion database name to view the list of tables.

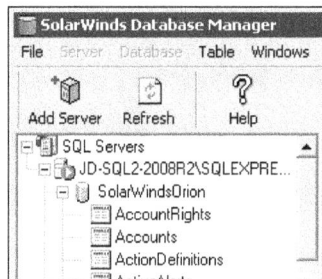

4. Scroll down to the **Engines** table. Right-click and choose **Query Table**.

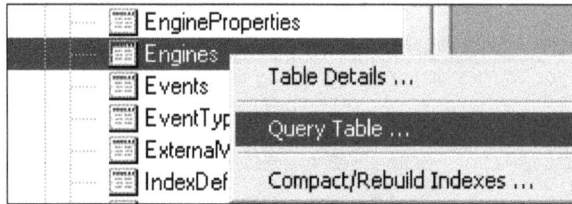

5. In the **Query** window, choose the **Read-Write** option and click on the **Refresh** button.

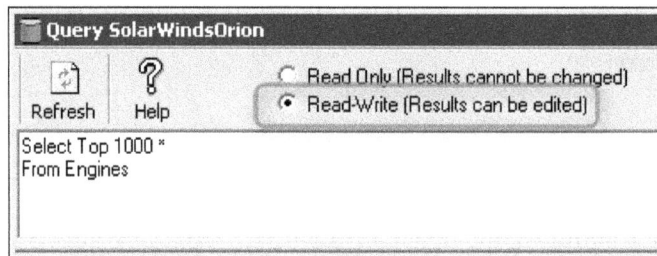

6. Each Orion NPM server will be listed in this display grid.

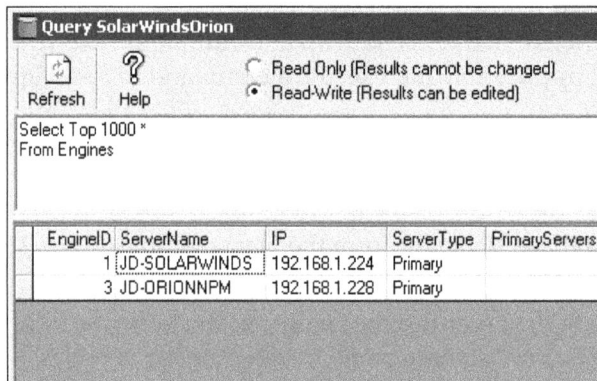

7. Swap each value in each of the **ServerName** field with one another. Also, update the IP addresses if needed. For example, my old Orion NPM server name was **JD-SOLARWINDS**. I deleted the **JD-SOLARWINDS** value and replaced it with the new server name **JD-ORIONNPM**.

8. When that is complete. Right-click on the second value and delete it.

EngineID	ServerName	IP	ServerType	PrimaryServers	
1	JD-ORIONNPM	192.168.1.228	Primary		11/15/20
3	JD-ORIONNPM	192.168.1.228	Primary		11/15/20
			Copy		
			Paste		
			Delete Selected Rows		

9. Close the **Database Manager** application and open **Orion Service Manager**. Start all of the Orion NPM services.

The new Orion NPM installation is up and running and all of the nodes that were being monitored by the old server are now being monitored by the new one.
To continue the migration, we need to copy the custom reports.

Copying custom reports

If you created any custom reports on the old Orion NPM server, you need to move them to the new Orion NPM installation and update the report schema. To do this, perform the following steps:

1. Close all open applications on the old Orion NPM server.

2. On the old Orion NPM server, copy the `Reports` folder to the same location on the new server.

   ○ Folder location in Windows Server 2003 is
     `C:\Program Files\SolarWinds\Orion\Reports`.

   ○ Folder location in Windows Server 2008 and above is
     `C:\Program Files (x86)\SolarWinds\Orion\Reports`.

3. After the reports have been copied, navigate to **Start | All Programs | SolarWinds Orion | Advanced Features | Custom Property Editor**.

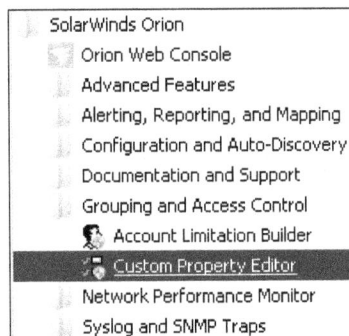

SolarWinds Orion
Orion Web Console
Advanced Features
Alerting, Reporting, and Mapping
Configuration and Auto-Discovery
Documentation and Support
Grouping and Access Control
Account Limitation Builder
Custom Property Editor
Network Performance Monitor
Syslog and SNMP Traps

4. Right-click on a blank area on the toolbar and choose **Customize**.

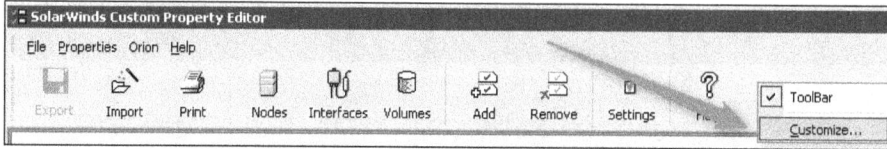

5. Highlight the **Properties** category, then drag **Update Report Schemas** to a blank area of the toolbar.

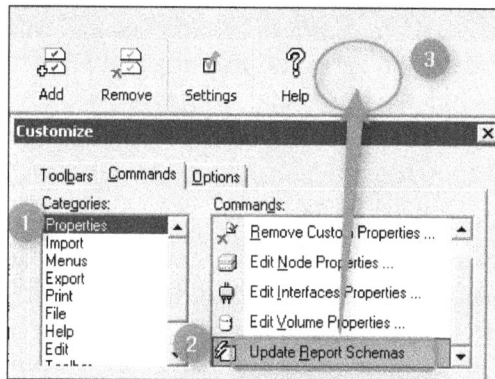

6. Close the **Customize** window and click on the **Update Report Schema** button in the toolbar.

7. After a moment, Orion will display a pop-up window notifying you if the task was successful or not. Click on **OK** to complete the update, then close the **Custom Property Editor** window.

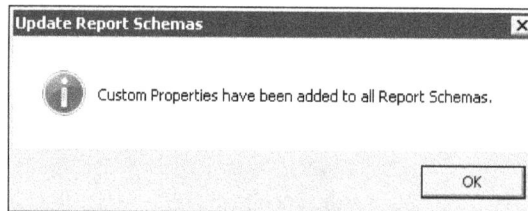

All of your custom reports as well as your custom properties will be restored on the new Orion NPM server. Simply open the **Reports** view in the Orion dashboard and run your reports as you always could!

Migrating Orion NPM licensing

Once you are comfortable with the setup on the new Orion NPM server, and before turning off the old Orion NPM server, you must deactivate the license from the old server in order to re-activate it on the new one. You have 30 days from the date you install the new Orion NPM server to migrate the license. This is the final task when performing a migration.

> **License Manager** needs to be installed on both the old Orion NPM server as well as the new Orion NPM server. Go back to the *License management* section of this chapter for guidance on how to install **License Manager**.

1. Log into the old Orion NPM server.
2. Launch the License Manager application on the old server by navigating to **Start** | **All Programs** | **SolarWinds** | **SolarWinds License Manager**.
3. Place a check mark next to the Orion NPM product title and click on **Deactivate**.
4. Specify your SolarWinds **Customer ID** and **Password** when prompted to do so, then click on **Deactivate** again.
5. Close the **License Manager** application and shut down the old Orion NPM server.
6. Verify that the license key from the old server is now available to be activated on the new one. To do this, log into the SolarWinds Customer Portal at `https://customerportal.solarwinds.com/customerportal/`.
7. Log into the new Orion NPM server.
8. Launch the **License Manager** application and activate the license key.

Deactivating and activating license keys can be done manually by an administrator without having to call SolarWinds customer support as long as each computer has access to the Internet and is able to communicate with SolarWinds' activation servers. If you are unable to re-activate the license key, you must call SolarWinds customer support for assistance.

Summary

You can see that, maintenance is a fairly important aspect of managing a SolarWinds Orion Network Performance Manager installation. I trust that you have found the information in this chapter extremely valuable, especially the information on migrating both an Orion database and an Orion NPM installation to new servers.

Documentation and Support
A

So, you have now reached the end of this book and have some questions about topics that were not covered or elaborated on and you need more information. I completely understand! Luckily, there are plenty of online resources for Orion NPM. SolarWinds provides a great deal of documentation on all of their products and you can find just about all of it through their main website at www.SolarWinds.com.

Documentation

The SolarWinds Orion documentation repository is located at http://www.solarwinds.com/documentation/orion/orionDoc.aspx and includes the following:

- Orion NPM datasheet
- Quick start guide
- Administrator manuals
- Reference guides

Knowledgebase

Another excellent resource is the SolarWinds knowledgebase. In it, there are thousands of how-tos, troubleshooting articles, FAQs, and information on best practices. If you encounter an error message or have a question about a topic covered in this book, the knowledgebase will likely hold an answer. To visit the knowledgebase, go to http://knowledgebase.solarwinds.com/kb/.

Training

SolarWinds provides free training in video form both on their official website, and through their YouTube channel. On www.SolarWinds.com, there are only a few basic customer training videos while the SolarWinds YouTube channel has over 50 videos just on Orion NPM! It doesn't hurt to take a look at some of these videos. Did I mention that they are free?

- Official Training: http://www.solarwinds.com/support/tutorials.aspx
- YouTube channel: http://www.youtube.com/solarwinds

Support

There may be times when you will need to contact SolarWinds technical support. Some examples of when you may need to contact support include the following:

- Assistance with an error message
- Licensing issues
- Information not available in the knowledgebase

You can find all of SolarWinds' contact information at http://www.solarwinds.com/support/.

B

The Thwack Community

Beyond the standard support and documentation resources that SolarWinds provides, there is also one more thing; the Thwack Community at www.thwack. com. Back in 2005, Thwack was an alternative for SolarWinds product support. There were only a few hundred members at the time and it was a simple forum that provided support and information for SolarWinds products. But today, Thwack is strong with more than 25,000 members and has a wealth of information not only about SolarWinds products but also on networking, server administration, and other IT professional topics.

The Thwack community has evolved from a simple forum in to a fully-featured IT professional community. In it, you can find blogs, forums, downloads, and product trial links, a request box for SolarWinds support, training videos, geek games, support articles, and advice, and above all else, the content exchange. SolarWinds product managers, developers, and SolarWinds Certified Professionals do frequent community posts on Thwack and respond directly to customers. In fact, many new product features come straight from customer suggestions on www.thwack.com! Joining the Thwack Community is free for everyone and all SolarWinds customers are encouraged to do so. The primary sections of Thwack include the following:

- **FORUMS & SPACES**
- **FREE TOOLS & TRIALS**
- **IDEAS & FEATURE REQUESTS**
- **CONTENT EXCHANGE**
- **LIBRARY & SUPPORT**
- **GROUPS**

FORUMS & SPACES

FORUMS & SPACES is one of the primary sections of Thwack. This section is an index of all of the various forum topics, blog sites, product forums, whiteboards, and bulletin boards for the entire Thwack community. You will find plenty of advice and guidance in the forums, as well as assistance from SolarWinds support teams.

FREE TOOLS & TRIALS

The **FREE TOOLS & TRIALS** page is essentially a marketing page for all of SolarWinds' software families, both paid and free. If you want to try out any of the Orion software products, this is the page that will help you find a download link. However, there are plenty of useful free tools available on this page such as the IP Address Tracker and the free TFTP Server.

IDEAS & FEATURE REQUESTS

The **IDEAS & FEATURE REQUESTS** area of the Thwack Community is where you can go to see what SolarWinds is working on for their next software release, and is the place to go if you have a feature request.

CONTENT EXCHANGE

Throughout this book, there were several areas where we discussed exporting your configurations, custom reports, scripts, universal device pollers, and templates from Orion NPM. The purpose of exporting your templates is not only for backups, but also for sharing. The **CONTENT EXCHANGE** page is SolarWinds' public online forum for sharing all of these exported files online. Sharing and downloading templates and content from the content exchange is completely free.

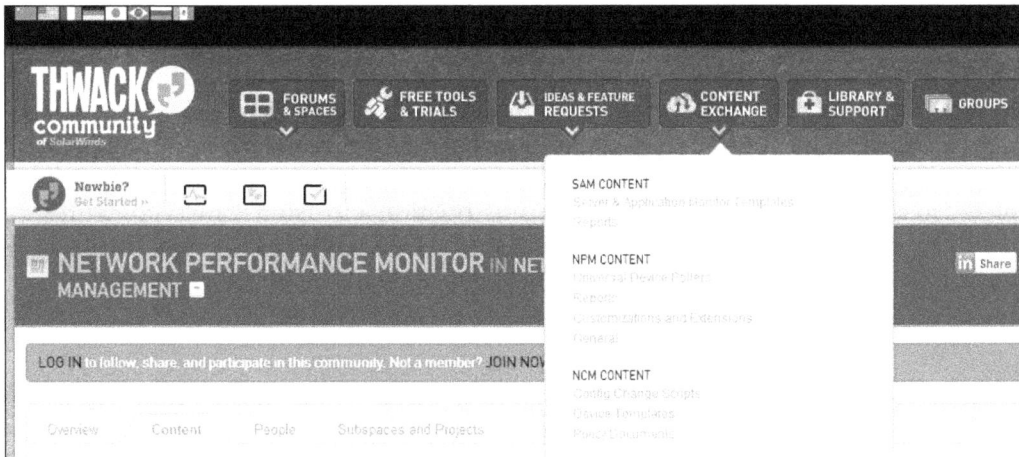

In the **CONTENT EXCHANGE** page, there is a separate forum area for all of SolarWinds' products where other customers post and share their own information and templates. I have personally used the content exchange to locate custom Universal Device Pollers for my APC power supplies, F5 Big-IP appliances, and more.

Several Thwack members have helped me accurately monitor UPS units, F5 BIG-IP appliances, and other network devices by sharing their Universal Device Pollers on the content exchange.

LIBRARY & SUPPORT

LIBRARY & SUPPORT is a one-stop-shop landing page that displays all of the relevant information for obtaining technical support from SolarWinds, and has links to all of the official documentation for Orion products. In addition there are links to some excellent training resources not only for Orion NPM, but also for CCNA, CompTIA Network+, and SolarWinds Certified Professional. The **LIBRARY & SUPPORT** page is definitely the one that you will want to bookmark in your web browser and visit from time to time.

GROUPS

The final area of the Thwack community is the **GROUPS** page. **GROUPS** is the social networking area of the Thwack community and is primarily an IT professional networking area. It is used as a place to share knowledge and learn from others. In it you will find groups that discuss a wide variety of things that may or may not be related to computers or networking including politics, television shows, computer repair, tutoring meetups, and more. However, the **GROUPS** page is also the place where you can sign up to be a part of SolarWinds' beta software programs.

C
Additional SolarWinds Orion Software

As you no doubt have seen throughout this book, SolarWinds built Orion NPM to be a robust product and it has the capability to monitor almost every aspect of your network. However, it is solely focused on monitoring your network equipment. What if you want to monitor NetFlow from your routers and switches? What if you need to be able to back up you configurations for your routers and switches? How about finding out how many Exchange server e-mail transactions are occurring, checking the latency of your web applications running in the cloud, and what if you need to be able to keep track of your public and private IP address space? To solve these questions, and many more, SolarWinds has several other products that extend Orion NPM's core functionality.

Each product is unique in what features they bring to the table. The additional SolarWinds Orion products are as follows:

- Network Configuration Manager
- Server & Application Monitor
- Patch Manager
- Web Performance Monitor
- IP Address Manager
- NetFlow Traffic Analyzer
- VoIP & Network Quality Manager
- User Device Tracker

All of the different SolarWinds products can be installed as their own instance or integrated as an add-on expansion to Orion NPM.

Network Configuration Manager

As more and more device configuration changes are made on your network, for example ACLs being added or changed on your firewalls, VLAN changes on your switches, or configuration changes to your routers, it becomes increasingly difficult to manage device configuration backups. Also consider this when you add more devices to your network, or when you replace devices. Keeping your devices backed up regularly, and centrally, can be a serious pain. This is where Orion **Network Configuration Manager** (**NCM**) can help.

Orion NCM is a management database, scripting engine, centralized deployment utility, and configuration backup system for all of your routers, switches, and firewall appliances on your network. Orion NCM can automatically scan the network for new or changed devices and automatically back up each device configuration, and can display the configuration history of each backed up device. You can use Orion NCM to compare device configurations side-by-side which allows you to easily revert back to a previous configuration and can mass-deploy commands to your network devices from a single pane of glass.

If any of the features listed features can assist you in managing your network, NCM is definitely worth the look. For more information about Network Configuration Manager, navigate to `http://www.solarwinds.com/network-configuration-manager.aspx`.

Server & Application Monitor

SolarWinds Orion Network Performance Monitor is an excellent tool for monitoring your network infrastructure. However, it does not have the ability to monitor applications, hardware health, and services. SolarWinds **Server & Application Monitor** (**SAM**) is an agentless application that can monitor the health of thousands of types of server hardware and server-based applications. Just as in Orion NPM, you can set up alerts and warnings for anything monitored by SAM and it can run as a standalone system or it can be integrated with Orion NPM. If you have the need to monitor server-based applications, operating system processes, and services such as Microsoft Exchange Server, SQL Server, Active Directory, or Oracle, definitely take a look at SAM. More information about SolarWinds SAM can be found at `http://www.solarwinds.com/server-application-monitor.aspx`.

Patch Manager

Are you familiar with Microsoft's **Windows Server Update Services (WSUS)**?
SolarWinds Patch Manager is the exact same thing in concept but has the ability
to do much more. Patch Manager allows you to not only manages Windows
updates for Windows Server and the Windows desktop, but also manages rolling
out software updates for a wide-array of third-party applications such as Adobe
Flash Player, Google Chrome, Mozilla Firefox, WinZip, and more.

Patch Manager has a robust reporting feature to help IT management
ensure compliance with all computers in the organization and it can be
extended to work with Microsoft System Center Configuration Manager.
Even though Patch Manager is designed to be integrated with Orion NPM,
Patch Manager can run as its own instance. More information, including all
of the third-party software updates that Patch Manager supports can be found
at `http://www.solarwinds.com/patch-manager.aspx`.

Web Performance Monitor

Web Performance Monitor (WPM) is a SolarWinds application that scans and
monitors websites hosted in Microsoft IIS, Apache, Tomcat, or any other web server.
Whatever the type of code, markup or libraries a website has been built with, from
HTML 5, jQuery, Twitter Bootstrap, or some other custom website code, WPM can
monitor it.

WPM traverses through web portals and web applications as if it is a user. It can
report, track, and alert against the speediness of each step taken while monitoring
your websites as well as record and playback each trip. If you need to track
the performance of your web services, it may be worth taking the time to look
at Web Performance Monitor. More information about WPM can be found
at `http://www.solarwinds.com/web-performance-monitor.aspx`.

IP Address Manager

IP Address Manager (IPAM) is a centralized IP address, DNS, and IP subnet
management system. It can track, report, alert, and manage your IP addresses.
Some of the core features of IPAM are IPv6 migration planning tools, historical
IP address tracking, subnet allocation wizard, and IP subnet scanning and alerting.

When IPAM is integrated with Orion NPM, all of its core features are directly integrated with Orion NPM and helps you manage your IP address space easily from the Orion dashboard. For more information on IPAM, go to `http://www.solarwinds.com/ip-address-manager.aspx`.

NetFlow Traffic Analyzer

SolarWinds **NetFlow Traffic Analyzer** (**NTA**) gives you a complete view of the bandwidth usage in your network in that it provides information about how the network is being used, who is using it, and for what purpose are they using the network.

NetFlow is a Cisco proprietary protocol. However, other vendors such as HP, Juniper, and Nortel Networks have their own versions of NetFlow called J-Flow, sFlow, IPFIX, and Huawei NetStream. SolarWinds NTA can analyze data from all of these types of flow protocols. More information about NTA can be found at `http://www.solarwinds.com/netflow-traffic-analyzer.aspx`.

VoIP & Network Quality Manager

Having a properly configured converged data network that services both voice and data can be difficult to troubleshoot when certain information is missing. For example, knowing that there is an issue with phone calls dropping off an IP phone due to a data network issue can be difficult to resolve. SolarWinds VoIP & Network Quality Manager can help.

VoIP & Network Quality Manager can monitor your organization's VoIP call performance, verify your IP service level agreement and Quality of Service policies, search and filter call detail records, and provide WAN performance alerts. More information about VoIP & Network Quality Manager can be found at `http://www.solarwinds.com/voip-network-quality-manager.aspx`.

User Device Tracker

Without creating some type of custom scripts or processes, it can be difficult to track down user devices to an access point, router, or switch port that the device is connected to. User Device Tracker helps to resolve this problem. It can quickly find computers, user names, MAC addresses, VLANs, and switch ports that devices are connected to. It can also show available ports on a switch and alert against the ones that may be getting close to operating at full capacity. You can also set up a device watch list by IP address, MAC address, or host name and receive an alert telling you what part of the network that device is connected to. More information about User Device Tracker is found at `http://www.solarwinds.com/user-device-tracker.aspx`.

Index

W

[PACKT] PUBLISHING enterprise
professional expertise distilled

Thank you for buying
SolarWinds Orion Network Performance Monitor

About Packt Publishing

Packt, pronounced 'packed', published its first book "Mastering phpMyAdmin for Effective MySQL Management" in April 2004 and subsequently continued to specialize in publishing highly focused books on specific technologies and solutions.

Our books and publications share the experiences of your fellow IT professionals in adapting and customizing today's systems, applications, and frameworks. Our solution based books give you the knowledge and power to customize the software and technologies you're using to get the job done. Packt books are more specific and less general than the IT books you have seen in the past. Our unique business model allows us to bring you more focused information, giving you more of what you need to know, and less of what you don't.

Packt is a modern, yet unique publishing company, which focuses on producing quality, cutting-edge books for communities of developers, administrators, and newbies alike. For more information, please visit our website: www.packtpub.com.

About Packt Enterprise

In 2010, Packt launched two new brands, Packt Enterprise and Packt Open Source, in order to continue its focus on specialization. This book is part of the Packt Enterprise brand, home to books published on enterprise software – software created by major vendors, including (but not limited to) IBM, Microsoft and Oracle, often for use in other corporations. Its titles will offer information relevant to a range of users of this software, including administrators, developers, architects, and end users.

Writing for Packt

We welcome all inquiries from people who are interested in authoring. Book proposals should be sent to author@packtpub.com. If your book idea is still at an early stage and you would like to discuss it first before writing a formal book proposal, contact us; one of our commissioning editors will get in touch with you.

We're not just looking for published authors; if you have strong technical skills but no writing experience, our experienced editors can help you develop a writing career, or simply get some additional reward for your expertise.

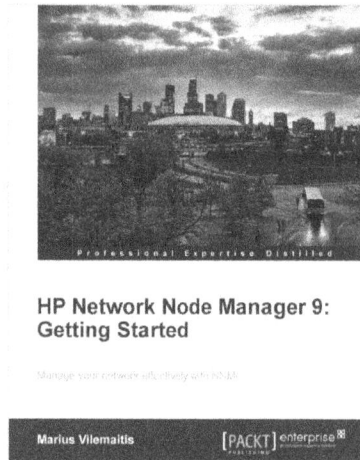

HP Network Node Manager 9: Getting Started

ISBN: 978-1-84968-084-4   Paperback: 584 pages

Manage your network effectively with NNMi

1. Install, customize, and expand NNMi functionality by developing custom features

2. Integrate NNMi with other management tools, such as HP SW Operations Manager, Network Automation, Cisco Works, Business Availability center, UCMDB, and many others

3. Navigate between incidents and maps to reduce troubleshooting time

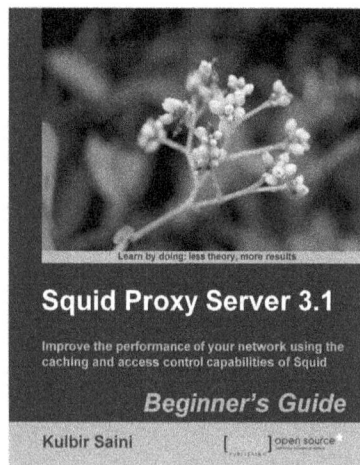

Squid Proxy Server 3.1: Beginner's Guide

ISBN: 78-1-84951-390-6   Paperback: 332 pages

Improve the performance of your network using the caching and access control capabilities of Squid

1. Get the most out of your network connection by customizing Squid's access control lists and helpers

2. Set up and configure Squid to get your website working quicker and more efficiently

3. No previous knowledge of Squid or proxy servers is required

Please check **www.PacktPub.com** for information on our titles

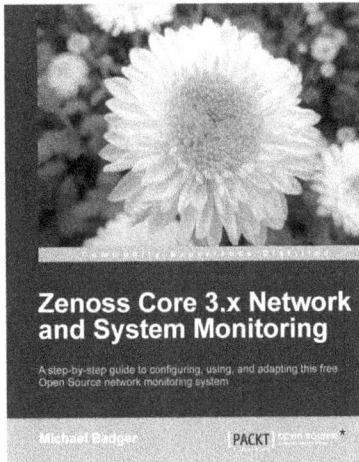

Zenoss Core 3.x Network and System Monitoring

ISBN: 978-1-84951-158-2   Paperback: 312 pages

A step-by-step guide for configuring, using, and adapting this free Open Source network monitoring system

1. Designed to quickly acquaint you with the core feature so you can customize Zenoss Core to your needs

2. Discover, manage, and monitor IT resources

3. Build custom event-processing and alerting rules

4. Write custom device reports to extract, display, and analyze monitoring data

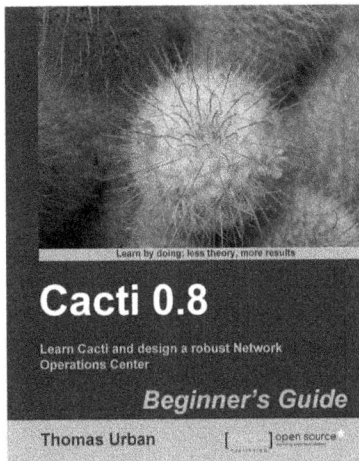

Cacti 0.8 Beginner's Guide

ISBN: 978-1-84951-392-0   Paperback: 348 pages

Learn Cacti and design a robust Network Operations Center

1. A complete Cacti book that focuses on the basics as well as the advanced concepts you need to know for implementing a Network Operations Center

2. A step-by-step Beginner's Guide with detailed instructions on how to create and implement custom plugins

3. Real-world examples, which you can explore and make modifications to as you go

Please check **www.PacktPub.com** for information on our titles

www.ingramcontent.com/pod-product-compliance
Lightning Source LLC
Chambersburg PA
CBHW082106220326

41598CB00066BA/5632